Gerontological Aspects of Genome Peptide Regulation

Vladimir Kh. Khavinson Vladimir V. Malinin

Gerontological Aspects of Genome Peptide Regulation

23 figures and 16 tables, 2005

KARGER

Basel · Freiburg · Paris · London · New York ·
Bangalore · Bangkok · Singapore · Tokyo · Sydney

••••••••••••••••••••••••

Vladimir Kh. Khavinson, MD
Director of the Saint Petersburg Institute of Bioregulation and Gerontology,
North-Western Branch of the Russian Academy of Medical Sciences,
Saint Petersburg, Russia.
Vice-President of the Gerontological Society of the Russian Academy of Sciences;
Associate Member of the Russian Academy of Medical Sciences.

Vladimir V. Malinin, MD
Head of the Laboratory of Peptide Pharmacology,
Saint Petersburg Institute of Bioregulation and Gerontology,
North-Western Branch of the Russian Academy of Medical Sciences,
Saint Petersburg, Russia.

Library of Congress Cataloging-in-Publication Data

Khavinson, V. Kh. (Vladimir Khatskelevich)
Gerontological aspects of genome peptide regulation / Vladimir Kh. Khavinson,
Vladimir V. Malinin.
p. ; cm.
Includes bibliographical references and index.
ISBN 3-8055-7895-4 (hard cover : alk. paper)
1. Peptides–Physiological effect. 2. Genetic regulation. 3. Aging–Molecular aspects.
[DNLM: 1. Gene Expression Regulation–physiology. 2. Aging. 3. Peptides–physiology.
QH 450 K45g 2005] I. Malinin, Vladimir V. II. Title.
QP552.P4K48 2005
612.6′7–dc22

2004030318

All options, conclusions, or regimens are those of the authors, and do not necessarily reflect the views of the publisher.

Bibliographic Indices. This publication is listed in bibliographic services.

Drug Dosage. The authors and the publisher have exerted every effort to ensure that drug selection and dosage set forth in this text are in accord with current recommendations and practice at the time of publication. However, in view of ongoing research, changes in government regulations, and the constant flow of information relating to drug therapy and drug reactions, the reader is urged to check the package insert for each drug for any change in indications and dosage and for added warnings and precautions. This is particularly important when the recommended agent is a new and/or infrequently employed drug.

www.karger.com
Printed in Switzerland on acid-free paper by Reinhardt Druck, Basel
ISBN 3–8055–7895–4

Contents

	Preface	VII
	Acknowledgements	IX
1	**Anti-Aging Mechanisms of Peptides**	1
2	**Estimation of Peptides Effect on Gene Expression in Mouse Heart Using DNA-Microarray Technology**	11
3	**Effect of Peptides on IL-2 Gene Expression**	21
3.1	Effect of Vilon and Epitalon on IL-2 Gene Expression in Mouse Spleen Lymphocytes	21
3.2	Effect of Vilon and Epitalon on IL-2 Gene Expression in Rat Hypothalamus Structures	24
4	**Effect of Peptides on *c-fos* Gene Expression in Stress**	30
4.1	Effect of Vilon on *c-fos* Gene Expression in Neurons of Rat Hypothalamic Paraventricular Nucleus under Emotional Stress	30
4.2	Effect of Vilon and Epitalon on *c-fos* Gene Expression in Cells of Different Hypothalamic Structures in Rats under 'Handling' Stress	31
4.3	Effect of Epitalon on c-Fos Protein Content in the Pineal Gland of Rats Exposed to Osmotic Stress	35
5	**Effect of Peptides on Carcinogenesis in Transgenic Mice**	38
6	**Effect of Peptides on Chromosomes**	40
6.1	Effect of Vilon and Epitalon on Number of Chromosome Aberrations in Corneal Epithelium and Bone Marrow Cells of Hypokinesic Rats	40

6.2 Effect of Epitalon on the Incidence of Chromosome Aberrations in Senescence-Accelerated Mice 41
6.3 Effect of Vilon and Epitalon on Nucleolus Organizers in Human Thymic Cells 43
6.4 Effects of Small Peptides on Blood Lymphocyte Chromatin in Senile Humans 46

7 Effect of Epitalon on Telomerase Activity, Telomere Elongation and Proliferative Potential in Human Somatic Cells 52

8 Effect of Peptides on DNA Synthesis 58
8.1 Effect of Vilon on DNA Synthesis in Rat Thymus Cells under the Influence of Nitrites 58
8.2 Effect of Vilon on DNA Synthesis in Blood Leukocytes of Old Rats Exposed to Small-Dose Radiochemical Impact 60
8.3 Effect of Vilon on Mitotic Activity of Hepatocytes in Rats after Partial Hepatectomy 62

9 Interaction between Small Peptides and DNA Double Helix 64
9.1 Peculiarities of Complementary Interaction in the 'Peptide-DNA' System 78
9.2 Model of Complementary Interaction between Small Peptides and DNA 79
9.3 Informational Interaction in the 'Peptide-DNA' System 86

10 From Anti-Aging Peptide Regulation to Pharmacogenomics 88

11 Conclusion 92

References 93

Subject Index 102

Preface

This monograph is the first work dedicated to the study of the interrelation between life span and peptide regulation of gene activity. We have developed a new concept which comprehensively reflects the evolutionary and biological role of peptides in the organism. Different chapters contain information on the effect of small peptides on the expression of the genes responsible for the regulation of metabolic processes, cell differentiation and proliferation, which provide for the maintenance of the main physiological functions and lead to a decelerated organism aging process. A wide range of physiologic effects of peptides, which is realized through the regulation of the expression of certain genes and through restoration of their structure is aimed at homeostasis support and delaying the realization of the aging genetic program. The mechanisms of the geroprotective effect of peptides, which are connected with chromatin activation, as well as with the increased enzymatic activity of telomerase and telomere elongation in different cells, have been discovered. Data reflecting the interaction between small peptides and DNA – the key point in the initiation of the biological activity of peptides – have been presented. Modern views reflecting the biological role of small peptides in the organism in humans and animals have been described, which consists of the genetic stability support aimed at increasing life span up to the specific limit.

A new approach to the study of aging mechanisms will resolve many controversial issues in anti-aging medicine. Further study of the molecular-genetic mechanisms of the effect of peptides may cause a reconsideration of the traditional views on aging, the creation of geroprotective means, as well as of accelerated aging and age-related pathology prevention methods.

This monograph may be of interest to gerontologists, specialists in genetics, molecular biologists, biochemists, pharmacologists, as well as clinicians.

Vladimir Kh. Khavinson
Vladimir V. Malinin

Acknowledgements

The authors express their sincere gratitude to the following well-known scientists and colleagues:

Prof. V.N. Anisimov, President of the Gerontological Society of the Russian Academy of Sciences; I.P. Ashmarin, Member of the Russian Academy of Medical Sciences, Head of Human and Animal Physiology Department, M.V. Lomonosov Moscow State University; K.R. Boheler, Investigator Head, Gerontological Research Centre, National Institute on Aging, USA; C. Franceschi, Research Director, Italian National Research Center on Aging, Italy; V.T. Ivanov, Member of the Russian Academy of Sciences, Director of the Institute of Bio-Organic Chemistry named after M.M. Shemyakin and Yu.A. Ovchinnikov, Russian Academy of Sciences; E.A. Korneva, Member of the Russian Academy of Medical Sciences, Head of the General Pathology and Pathophysiology Department, Research Institute of Experimental Medicine, Russian Academy of Medical Sciences (RAMS); Prof. T.A. Lezhava, Head of the Genetics Department, Tbilisi State University, Georgia; Prof. J. Martinez, Director, Laboratory of Amino Acids, Peptides and Proteins, University of Montpellier, France; Prof. M. Passeri, Past President of the International Association on Gerontology – European Region, Italy; G.A. Sofronov, Member of the Russian Academy of Medical Sciences, Scientific Secretary of the Academic Council, Military Medical Academy; K.V. Sudakov, Member of the Russian Academy of Medical Sciences, Director of the Research Institute of Normal Physiology named after P.K. Anokhin, RAMS; Prof. A.I. Yashin, Center for Demographic Studies, Duke University, USA; and to the researchers and officials of the Saint Petersburg Institute of Bioregulation and Gerontology of the North-Western Branch of the Russian Academy of Medical Sciences:

Dr. S.V. Anisimov; Dr. I.E. Bondarev; Prof. I.M. Kvetnoy, Dr. O.N. Mikhailova; Prof. V.G. Morozov; Dr. G.A. Ryzhak; Prof. L.K. Shataeva for useful advice and for their significant contribution into our research work on this problem.

1

Anti-Aging Mechanisms of Peptides

During the last decade, achievements in theoretical and practical gerontology have made up a foundation for the targeted regulation of age-related changes. Hence, one of the main tasks for modern gerontology is the prevention of premature aging and age-related pathology, aimed at increasing mean life span, maintaining active longevity and reaching the specific limit of human life span [Hayflick, 2000; Andrews, 2001; Gutman, 2002].

It is well known that aging is characterized by an intricate complex of molecular-genetic and biochemical changes, which are accompanied by a disorganized peptidergic system of organism functions regulation. The study of age-related involution of the organs and tissues of the organism showed a decrease in production of physiologically active substances of peptide nature, as well as in protein synthesis intensity, which enabled us to conclude that peptides play an important role in the regulation of aging mechanisms. The system of peptides is regarded as universal in the case of neuro-immune-endocrine interactions [Ashmarin and Korolyova, 2003]. Along with that, peptide regulation of the organism's physiological functions involves tissue-specific peptides supporting cellular homeostasis. Peptide bioregulators contained in different cells and tissues are produced in the process of limited proteolysis, and exert a wide array of biological effects coordinating the processes of development and functioning of multicellular systems. Despite their multilevel hierarchy, the peptide homeostasis regulation mechanisms perform a common task, i.e. they coordinate biosynthesis processes in organism cells by means of impact on gene expression.

Peptide regulation of homeostasis dominates in a complex chain of processes leading to the aging of cells, tissues, organs and the organism as a whole [Khavinson, 2002; Anisimov, 2003]. The involution of organs and tissues, first of all those related to the main regulatory systems – nervous, endocrine and immune – is a morpho-functional manifestation of aging [Petrov and Khaitov, 1975; Korneva and Schkhinek, 1988]. There are data to evidence the development of age-related changes in the pineal gland, thymus, as well as in brain cortex neurons and subcortical structures, retina, vascular wall and genital organs [Khelimsky, 1969; Chazov and Isachenkov, 1974; Hirokawa, 1980; Arking, 1998].

Involutionary changes in organs and tissues in the course of aging lead to weakening of their functions, the manifestation of which on molecular level consists prevalently in disturbed synthesis of specific proteins and altered gene expression level.

Age-related thymus involution is characterized by the most pronounced, as compared to other organs, structural and functional changes, similar in different kinds of mammals. In humans, this process begins already during the first year of life [Bodey and Kaiser, 1997], and, according to certain data, even before birth [Pawelec et al., 1999]. In the course of thymus involution, a decrease in the quantity of T lymphocytes is registered in the organism, and B lymphocyte differentiation into antibody-producing plasma cells is infringed. It is presumed that a critical role in this process is played by the disturbed regulatory functions of T cells. Besides thymus involution, the following factors also contribute to the reduced immune function of T cells: stem cell defects, defects in antigen-producing cells, aging of resting immune cells, disturbed immune cell activation, replicating depletion of proliferating cells clones [Pawelec et al., 1999]. These changes affect the whole organism. In particular, the age-related immune system exhaustion contributes to the increase in the incidence of malignant tumors, the decrease in resistance against infections and an increased frequency of autoimmune disturbances, as well as a shift in the scope of produced interleukins towards those contributing to inflammatory reactions [Ouyang et al., 2000].

One of the mechanisms, by means of which thymus involution affects organism status is a decrease in the production of thymus peptide factors. As an example of regulatory peptides acting as biological markers of aging one can name thymulin, thymus peptide (thymus serum factor), which consists of 9 amino acids. Beginning approximately from the age of 20 years, its content decreases until it virtually disappears by the age of 50–60 years. Some food factors (Zn, vitamin E) may prolong this period. The level of these two important regulatory peptides – α_1-thymosin and thymopoietin – begins to rapidly decrease by the age of 10 years and by the age of 30–40 years they can hardly be identified [Chakravarti and Abraham, 1999]. There are certain reasons to assume that, except for thymulin, certain immune regulatory functions are performed by some other regulatory peptides, still unknown or insufficiently studied, from the thymus where oligopeptides, both identified and unidentified, are present in abundance.

The pineal gland can be an example of the organ in which, unlike the thymus, no radical structural changes are observed in the course of aging. Meanwhile, the age-related functional changes in the pineal gland are quite significant, and they, as in the thymus, exert a drastic influence on the whole organism. Being a neuroendocrine organ, the pineal gland (epiphysis) is considered to be one of the key interrelation points of age-related changes, which became clear in the course of accumulation of the data reflecting the aging process in mammals [Reiter, 1995]. Pineal gland hormone – melatonin (N–acetyl-5-metoxytriptamine) – possesses a unique scope of effects. Pineal gland melatonin production has a pronounced 24-hour cyclic character with night-time peaks [Dawson and van den Heuvel, 1998]. In the course of aging, the night-time melatonin secretion peaks are reduced, which causes numerous consequences for the neuroendocrine system, mainly due to reproductive functions and stress responses, as well as for the immune system, for carcinogenesis and for wake/sleep cycle regulation [Reiter, 1995].

Many known regulatory peptides or signs of their presence have been found in the pineal gland [Moller et al., 1996; Debeljuk and Lasaga, 1999]. Part of these peptides is secreted by axon endings from other brain structures. Y neuropeptide (NPY) and other tachykinins (A neurokinin, K neuropeptide) are contained in the axons at the upper cervical ganglia. P substance is provided by the trifacial nerve ganglion. Nerve fibers with immune reactivity for PACAP (pituitary adenylate cyclase-activating peptide) as well as oxytocin and secretoneurin have been found in the pineal gland. Somatostatin is produced in the pineal gland itself and in interneurons. Some pinealocytes contain enkephalin, their outgrowths ending on enkephalin-negative pinealocytes. Endorphin is present in other pinealocytes. Pinealocytes also possess membrane receptors for opioids. Immune reactivity for arginine-vasopressin is also present in pinealocytes, as well as gonadotropin-release factor (GnRF) [Moujir et al., 1990; Park et al., 1995]. Judging by the presence of mRNA, arginine-vasotocin is expressed in the pineal gland [Badiu et al., 1999], together with type C natriuretic peptide [Middendorff et al., 1996]. Nerve growth factor has been found in glial pineal gland cells [McNulty et al., 1995]. The main fibroblast growth factor has been found in the cells, which are characterized as ‘not pinealocytes’, as well as in endothelium [Marin et al., 1994].

The situation concerning the endocrine secretion of pineal gland peptides remains unclear. It is presumed that arginine-vasotocin is one of the possible pineal gland endocrine products of peptide nature [Goldstein, 1992], but the corresponding mRNA has only been found in a few pineal gland cells [Badiu et al., 1999]. Thr-Ser-Lys tripeptide, isolated from the pineal gland, inhibits puberty in rats and is a fragment of antigonadotrophic decapeptide [Benson et al., 1996]. It is assumed that the oncostatic activity of the pineal gland is mediated not only by melatonin, but also by the secretion of some factors with 0.5–1 kDa molecular mass, which are present in the extract from the pineal gland which contains no melatonin [Bartsch et al., 2000].

Among other peptide factors which have been found in the pineal gland, only vasopressin has been studied in respect to aging. In rats, the content of this peptide in the pineal gland is not reduced, unlike that in hypothalamus, thalamus, midbrain and tonsil [Terwel et al., 1992].

Fundamentally important is the fact that the synthesis of regulatory peptides in cells is reduced in the course of aging, and that the sensitivity of target cells for them is changed [Kaskow et al., 1999]. Taking into consideration the wide range of biological activities of these substances, we can assume that their reduced production disturbs peptide regulation mechanisms and leads to the gradual fading away of functions in the senescent organism.

It is noteworthy that peptides can take part in hormonal, neurocrine, paracrine and autocrine regulation mechanisms. So, for example, hormonal regulation is characteristic for releasing hormones of hypothalamus. It was shown that in elderly men the hypothalamus produces less gonadotrophin-releasing hormone (luliberin), and Leydig cells in the testis give a weaker response to pituitary gland gonadotrophin [Mulligan et al., 1999]. In old rats, the response of neurosecretory hypothalamus cells, synthesizing luliberin, to NMDA receptors excitation, is weakened [Bonavera et al., 1998]. The result of such peptide regulation distuurbances is the age-related weakening of luteinizing hormone secretion by adenohypophysis cells, which leads to hormonal disorders [Sahu and Kalra, 1998]. A low level of mRNA corticotrophin-releasing hormone was been detected in adenohypophysis of old rats [Kaskow et al., 1999], alongside with a reduced concentration of arginine-vasopressin in the suprachiasmatic hypothalamic nucleus [Van der Zee et al., 1999] as well as in blood [Mulchahey et al., 1999]. Meanwhile, a partial decrease in the expression of mRNA of vasoactive intestinal peptide (VIP) was observed in the neurons of the suprachiasmatic nucleus in old animals, while mRNA levels of arginine-vasopressin and somatostatin remained unchanged [Duncan et al., 2001].

Manifestations of aging are considered to be associated with a physiological decrease in growth hormone level. The analysis of age-related changes of gene expression in rat liver showed that the largest group of age transcripts coded the proteins involved in intermediate metabolism, chondriosome width and medical substance metabolism. Approximately 40% of expressed gene products were normalized after growth hormone administration [Tollet-Egnell et al., 2001].

It was proved that postnatal mice showed a decreased level of expression of natriuretic C-type peptide mRNA in different organs and tissues [Stepan et al., 2000].

For the first time it was found that the difference in growth factor gene and protein expression between dorsal and ventral lobes of the prostate is related to prostate aging, as well as to pathogenesis of age-related prostate diseases [Zhao et al., 2002].

As already stated, the reduction of peptide synthesis level is accompanied by the weakening of these peptides' perception by receptors of their target cells. So, for example, the effect of exogenously administered P substance on the neurons of solitary tract nucleus in old rats is much less pronounced than in young animals. Obviously, this is connected with the revealed decrease in NK-1 receptors quantity (specific for P substance) in this region of the brain [Mazzone and Geraghty, 1999]. It was also shown that the efficacy of P substance activity on respiratory tract smooth muscles is weakened. An important role in these phenomena may be played by the fact of age-related decrease in mRNA, which is necessary for the synthesis of the P substance predecessor preprotachykinin A [Mhanna et al., 1999]. In old mice, the stimulating effect of bombazine, gastrin-releasing peptide and active cholecystokinine fragment (CCK8 octapeptide) on lymphocyte chemotaxis is reduced or totally disappears. Moreover, in the course of aging the gastrin-releasing peptide and CCK8 octapeptide give a weaker stimulus to cytotoxic activity of natural killers [Medina et al., 1998]. In the course of aging, the level of VIP in ganglion stellatum is reduced [Roudenok et al., 1999]. The response of smooth muscles of the small intestine wall to VIP effect is also weakened [Erlwanger et al., 1999].

The analysis of peptide growth factor dynamics in the course of aging reveals polysemantic changes. So, synthesis of the fibroblasts' main growth factor necessary for the repair of neurons and their outgrowths is subject to a significant age-related decrease [Kuzis et al., 1999], while the level of thrombocyte growth factor increases. Aging is accompanied by an increase in the level of tumor necrosis factor production [Pekary and Hershman, 1998].

These alterations in the senescent organism may be either the consequence of a general imbalance of regulatory factors or a manifestation of some age-related changes compensation processes. Such disorganization in the production of peptide growth factors may also lead to numerous negative consequences.

Studies of age-related protein production dynamics in the organism virtually always show that in the course of aging both the speed of protein production growth under the effect of inductor and its maximum attainable level are reduced. This means that the real production of every protein would be decreasing in the course of aging, if regulatory factors, stimulating and inhibiting its synthesis, maintained a constant ratio. However, due to the existence of many regulatory links both in the organism as a whole and in every one its cells, the real levels of different protein generation (gene expression) against the background of lowering potentials can change in different directions and to a different extent [Golubev, 1989]. For example, the maximum pituitary gland capacity for peptide hormone production, in particular follicle-stimulating growth hormones (FSG), is reduced in the course of aging, but the production of sexual steroids inhibiting gonadotrophin secretion is also reduced, which causes a 4- to 5-fold increase in real FSG production beginning from puberty and up to the age of 60 years.

Different variants of age-related changes in protein production, identified according to these proteins' genes expression, are demonstrated in the studies with several thousands of mRNA units simultaneously registered by means of DNA microarrays [Cao et al., 2001; Kayo et al., 2001; Weindruch et al., 2002]. A 2-fold and even higher increase of expression was observed in 300 of 4,677 identifiable genes (4.2%) of macaque skeleton muscle in the course of aging, while a 2-fold and over decrease in the level of expression was registered in 2.1% of the genes. In mouse liver, the number of known genes, the expression of which changed in the course of aging, made 0.9% of all 11,000 identifiable genes, the ratio of decrease and increase cases was approximately equal: 57% and 43%, respectively. Among the genes with increased expression those were prevalent, the protein products of which took part in inflammatory reactions. The same was registered in the study of muscles in monkeys and mice, as well as of brain cortex and cerebellum of mice. Among the 26 genes identified in the liver, the expression of which decreases during the course of aging, 6 genes participate in the inhibition of cell growth and proliferation, confirming a weaker control over these processes, thus enhancing the risk of tumor growth. Among other genes with decreased expression are the genes defining specific liver functions. In skeleton muscles, an age-related decrease was detected in the activity of genes, the products of which take part in energy metabolism, and in particular in electron transportation through mitochondrial width chains and in oxidative phosphorylation, which is directly related to the specific functions of the muscles. The increased expression was also found in the genes, the products of which are involved in antioxidant functions, which points out the necessity of stronger resistance to free radical processes. Food caloric value restriction retarding aging in experimental animals is accompanied by the gene-expression profile in old animals approximated to that of young animals.

Thus, the obtained results confirm that the condition of an old organism can be approximated to that of a young organism, by resisting age-related disorders of the immune system, as well as tumor growth and oxidative stress, and by stimulating tissue-specific functions. At the same time, the above-described studies of genes, the expression of which changes in the course of aging, give little evidence of the theory that the production of regulatory peptides is subject to drastic age-related changes. Regarding this fact, one must consider the following. Firstly, the tissues not considered as main regulatory peptide generators in the organism and, moreover, themselves depending on such peptides secreted by other organs were studied. In particular, the main cause of senile muscular dystrophy (sarcopenia) is assumed to be a decreased production of peptide growth hormone by the pituitary gland, which in turn can be due to insufficient dopaminergic stimulation of growth hormone production [Golubev, 1989]. Secondly, the production and effect of regulatory peptides depends not only on their genes expression, but also on the synthesis of corresponding proteins in ribosomes, as well as on protein processing and their targets condition. A total age-related decrease in protein biosynthesis is, most probably, due to reduced translation processes [Tavernarakis and Driscoll, 2002]. Against this background, the consequences of altered specific gene expression, which in some cases may lead to age-related increases in protein production, are manifested. The next level of changes is protein concentration also depending on its processing. For example, age-related disturbances in opioidergic mechanisms of the central regulation of reproductive functions are related to the changes occurring at the convertase level and entailing the accumulation of inactive β-endorphin in the preoptic hypothalamic nucleus [Miller et al., 1998].

Thus, the analysis of age-related changes in homeostasis peptide regulation evidence a common tendency – aging is accompanied by gene expression alteration, reduced synthesis and secretion of regulatory peptides and by decreased target cell sensitivity to them.

Taking into consideration the important physiological role of endogen peptides in the regulation of the aging process, the development and study of geroprotective means of peptide nature is deemed to have good prospects. The urgency of development of effective geroprotective means is evident due to the evergrowing scope of adverse effects on the human organism and to the premature development of age-related pathology. The peculiarity of adaptive transformations under the effect of factors causing accelerated aging is that these transformations occur against the background of age-related changes in neuro-humoral regulation and the reduced synthesis of tissue-specific proteins; this finally results in reduced reserves of the organism, catabolism product accumulation, progressing destructive processes, disturbed cell functions and development of diseases [Dilman, 1986, 1994]. Despite the wide array of theories on aging, there is a logic interconnection between the effect of all identified factors, launching a step-by-step genetic program of cell degeneration. The leading role in this process belongs to the altered level of expression of different genes regulating the processes of cell proliferation, differentiation and death. All this confirms the stochastic character of these phenomena and the importance of the individual resistance of both separate cells and an organism as a whole to different stressor effects.

The first studies, the results of which formed a basis for the new concept of organism function peptide regulation, were performed by Morozov and Khavinson (1971–1975). These authors developed and implemented the original method of endogen peptide extraction from animal organs and tissues. In 1981–1983, a concept was put forward regarding the role of the new class of peptide bioregulators, which became generally known as 'cytomedins', in the support of structural and functional homeostasis of cell populations [Morozov and Khavinson, 1983].

Cytomedins were found to be generated in cells, and in terms of composition they are peptide complexes with molecular masses of 1–10 kDa. According to the proposed concept, they are able to perform trans-membrane transition of information necessary for development, interaction and functioning of cells. In other words, endogen peptide employment by cells may be regarded as one of the methods of the cell populations' self-regulation and, consequently, of cellular homeostasis [Kuznik et al., 1998].

While studying the biologic properties of extracted peptide complexes, it was revealed that these factors play a direct role in the processes of tissue-specific regulation of gene expression and biosynthesis. As a result of peptide regulation, the speed of pathologic changes accumulation in the cells is decreased (DNA damage, mutations, malignant transformation, etc.) and the activity of repair processes aimed at cell homeostasis restoration is increased.

Numerous studies revealed endogen peptide participation in the regulatory processes of both cell differentiation and proliferation, thus changing the functional activity of the genome, as well as the processes of protein synthesis with respect to the status of multicellular system. The role and origin of peptide bioregulators become clear if life manifestations are considered as an evolutionary process of genetic information exchange and reproduction. As we know, during the early stages of multicellular system development the leading role in the processes of gene expression regulation belongs to the cell mediators. That is why mediator mechanisms of cell self-regulation are the first ones to appear, and they do not reveal special specificity, but do reveal specificity, which enables them to control the mechanisms of multicellular organism development. Mediator systems, formed during the early stages of evolution, are assumed not to lose their role even after specialization of cells, giving them an important mechanism of gene-differentiated activity regulation.

Disturbed peptide regulation reduces the resistance of cells and tissues of the organism to destabilizing factors of both external and internal environment. This may serve as one of the causes for the development of numerous diseases, for involution of organs and tissues, as well as for premature aging.

The administration of peptide bioregulators under conditions of disturbed cellular homeostasis enables to restore functional activity of corresponding physiological systems and organs. At present, regulatory peptide complexes have been extracted from practically all cells, tissues and biological liquids of the organism. According to the data received from physical and chemical studies, they differ in terms of composition, molecular mass and electrochemical properties of their components.

The creation of medicinal substances on the basis of physiologically active endogen peptides produced by the organism, enabled to develop and introduce into clinical practice a new technology aimed at correction of genetic predisposition of the organism to age-related pathology and at the restoration of cell functions lost in the course of aging – i.e. the bioregulative therapy. It is based on the complex administration of endogen cell metabolism bioregulators for prophylactic and therapeutic purposes both extracted from animal organs and tissues and synthesized ones [Morozov and Khavinson, 1996; Khavinson and Morozov, 2001; Khavinson, 2002]. It is noteworthy that periodical administration of peptide bioregulators extracted from thymus (thymalin) and pineal gland (epithalamin) to patients over 60 years, resulted in partial restoration of functions of immune and cardiovascular systems, brain, liver and pancreas. This effect of peptides contributed to a reduced incidence of different disease exacerbations and was accompanied by a 2-fold decrease in death rate of patients during the follow-up period (6–8 years) [Korkushko et al., 2002; Khavinson and Morozov, 2003]. On the whole, bioregulation therapy may be regarded as a homeostatic branch of gerontology revealing large prospects.

It was found that the administration of complex polypeptide preparations (cytomedins) extracted from thymus, pineal gland, prostate, brain cortex and retina leads to significant restoration of protein synthesis in these organs and tissues, and to normalization of their functional activity in case of different pathologic states or aging. The study of structural and functional peculiarities of these substances is of principal importance for further understanding of the mechanisms of the geroprotective activity of peptides. It is necessary to note that among known peptides revealing geroprotective properties a special place is occupied by dipeptide carnosine (β-Ala-His), which was the first to be discovered among all the endogen substances of peptide nature. It was shown that the geroprotective effect of carnosine is connected predominantly with antioxidant activity of the peptide [Boldyrev, 1998; Hipkiss et al., 2001]. Amino acid analysis of natural polypeptides prompted the synthesis of peptides showing a geroprotective effect: Vilon (Lys-Glu) [Khavinson et al., 2003d], Epitalon (Ala-Glu-Asp-Gly) [Khavinson, 2004], Cortagen (Ala-Glu-Asp-Pro) [Khavinson et al., 2003c], Livagen (Lys-Glu-Asp-Ala) and Prostamax (Lys-Glu-Asp-Pro) [Khavinson and Malinin, 2002]. Comparative study of the biological activity of polypeptide preparations and synthetic peptides revealed a number of similar effects regarding their effect on different normal organs and tissues of the organism and concerning their pathology [Khavinson, 2002].

The results of long-term studies suggest that the geroprotective effect of peptides is related to the regulation of gene expression and the biosynthesis process, which is disturbed in the course of aging. This is even more important if we take into consideration the determining role of the genetic apparatus in the mechanisms of individual development and occurrence of many diseases [Bochkov, 1999; Korochkin and Mikhailov, 2000]. It is important to note that regulatory peptides are medicinal preparations of a new generation and that they perform naturally directed corrective action without side effects [Petrov et al., 2000].

Thus, the results of experimental and clinical study of peptide preparations evidence geroprotective properties of these substances, stipulated by their normalizing effect on basic systems of organism homeostasis support and by the regulation of aging mechanisms [Khavinson, 2002]. Peptides contribute to the restoration of protein synthesis intensity and regulate the expression of the genes responsible for cell differentiation and proliferation. Peptides may be also regarded as information regulators of genetic stability. This leads to the stabilization of basic physiological functions and inhibition of the aging process.

Basing on the above facts, we can assume that *aging is an evolutionary determined biological process of age-related change in the expression and structure of genes, which leads to the infringed synthesis of regulatory tissue-specific peptides in different organs and tissues followed by their structural and functional changes and development of diseases.*

This monograph contains basic results obtained during the study of molecular-genetic mechanisms of the action of small peptides, revealing new prospects in the development of the concept of peptide regulation of aging, as well as in the prevention of premature aging, age-related pathology and in prolonging the period of active human longevity.

2

Estimation of Peptides Effect on Gene Expression in Mouse Heart Using DNA-Microarray Technology

In order to estimate the impact of Vilon (Lys-Glu) and Epitalon (Ala-Glu-Asp-Gly) on gene expression, an experiment based on DNA-microarray technology was conducted. The modification of this method used during this experiment allows estimating the changes in the expression of 15,247 genes. Using this method, levels of mRNA in mouse heart before and after Vilon and Epitalon administration were studied [Anisimov et al., 2002].

Experiments were performed on 30 female CBA mice receiving standard laboratory food and water ad libitum. Six-month-old mice were randomly divided into 3 groups (10 animals in each group). Experimental mice received subcutaneous injections of Vilon and Epitalon (1 μg) in 0.1 ml of 0.9% NaCl for 5 days. Control mice were injected with 0.1 ml of 0.9% NaCl. The animals were decapitated on day 6. The hearts were removed, immediately frozen in liquid nitrogen, and stored at −80°C for isolation of total RNA. We used 15,247 clones from a cDNA library (National Institute on Aging, USA, NIA mouse 15 K cDNA clone set). Clones in the pSPORT-1 vector were grown in *Escherichia coli* using the nutrient broth LB/ampicillin. Plasmid DNA was isolated using Edge Biosystems kits (USA). cDNA inserts were amplified on a PCT-225 Peltier amplifier (MJ Research) using the direct primer 5′-CCA GTC ACG ACG TTG TAA AAC GAG-3′ and reverse primer 5′-GTG TGG AAT TGT GAG CGG ATA ACA ACA A-3′. The images were formatted using ImageTools 2.1 software and analyzed by means of ImageQuant 5.1 software (Molecular Dynamics).

In the course of the experiment, a triple hybridization was conducted of the microarrays with 15,247 cDNA clones in heart tissue samples from control and experimental mice. Variability of the results in control mice was lower than in

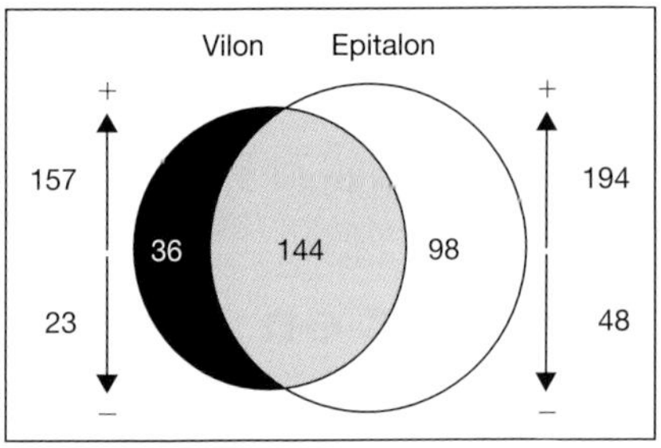

Fig. 1. Effect of Vilon and Epitalon on gene expression in mouse heart (Eiler-Wenn diagram).

experimental animals. After 3 hybridizations, the variation coefficients in control animals and mice receiving Vilon and Epitalon were 14.9–16.5, 16.1–16.6, and 18.9–19.3%, respectively. The mean correlation coefficients in these animals were 0.935, 0.976 and 0.969, respectively, which confirms the efficiency of hybridization of independently synthesized radioactive samples with microarrays was highly reproducible. The results of hybridization of microarrays with 15,247 cDNA clones in heart samples from control animals were compared with those of the mice treated with Vilon and Epitalon for 5 days. It was revealed that the expression of 300 clones (1.94% of the total count) in heart samples from mice treated with Vilon and Epitalon changed more than twice as compared to the control. Building of an Eiler-Venn diagram (fig. 1) shows that the administration of Vilon changed the expression of 36 clones, while Epitalon modulated the expression of 98 clones. Combined treatment with Vilon and Epitalon changed the expression of 144 clones. These results show that Vilon and Epitalon changed the expression of 180 and 242 clones, respectively. Vilon alone or in combination with Epitalon activated the expression of 157 clones (maximum 6.13 times), but inhibited the expression of 194 clones (maximum 2.79 times). Epitalon activated the expression of 194 clones (maximally 6.61 times), but inhibited the expression of 48 clones (maximally 2.71 times).

In 7 measurements, 2–15 of 300 clones corresponded to the same genes. Overall, 300 clones were associated with 278 genes. Some clones were similar to mitochondrial (n = 5, 1.8%) and nuclear mouse genes (n = 83, 29,9%). Other clones showed strong similarity to the nucleotide sequences of genes in humans (n = 51, 18.3%) and other organisms (n = 5, 1.8%). 134 clones (48.2%) revealed no similarity to any genes in humans or other organisms.

In general, nucleotide sequences of 144 clones were similar to genes in mice and other organisms, including *Homo sapiens, Rattus norvegicus, Macaca fascicularis*, and *Bos taurus*. Vilon and Epitalon changed expression in 5 of 13 genes of the mitochondrial genome. Expression of 4 mitochondrial genes (16S, Vilon and Epitalon; NADH dehydrogenases 4 and 5 and cytochrome b, Epitalon) increased 2.03–6.61 times. However, Vilon decreased the intensity of expression for ATPase 6 genes 2.25 times. According to functional classification of genes in the cardiovascular system [Hwang et al., 1997] 139 nuclear genes were divided into several functional categories. Some genes (n = 44) did not belong to any category or encoded hypothetical proteins. Other genes (n = 95) represented 6 main functional categories, including genes for cell division (n = 14), cell signaling systems and communication (n = 15), cell structure and motility (n = 7), protective systems of cells and organisms (n = 16), expression of genes and proteins (n = 24), and metabolism (n = 19) (table 1).

Genes, whose expression changed under effect of Vilon and Epitalon, functionally belong to various cell systems. It should be emphasized that the ratio of genes belonging to various functional categories and subcategories in the heart did not correspond to normal [Hwang et al., 1997]. The ratio of genes for expression of genes and proteins, including posttranscriptional factors (10 genes) and metabolism was the highest. Categories of genes for cell signaling systems, communication, and cell structure and motility did not differ from the control. However, categories of genes for cell division (10.07 vs. 5.68% in the control) and protective systems of cells and organisms (11.51 vs. 6.71% in the control) included a higher number of genes with modified expression. The data suggest that genes of these subclasses are most potent effectors for the biological effect of Vilon and Epitalon. Detailed studies of functional subcategories showed that changes in the ratio of genes for cell cycle regulation (subcategory 1d, 5 genes, 3.6% vs. 1.59% in the control) and protective systems of cells and organisms, including membrane transport/transport proteins (subcategory 4b3, 6 genes, 4.32 vs. 0.9% in the control) and immune system (subcategory 4c, 5 genes, 3.6 vs. 1.75% in the control) are of considerable importance. The effects of Vilon and Epitalon were compared on genes in heart samples. Epitalon selectively modulated the expression of genes associated with ribosomal proteins (subcategory 5b3). Vilon activated the expression of only 1 gene belonging to this subcategory (ribosomal protein S6 kinase 1, p70/p85 s6 kinase). Epitalon activated the expression not only of the s6 kinase gene, but also of genes encoding ribosomal proteins l8, L27, S6 and S10 (by 2.02–2.46 times). The gene for ribosomal protein S10 was presented by 2 clones which had similar nucleotide sequences to rat and human genes (table 1, shows only the human gene). Expression of these genes increased by 2.02 and 2.09 times, respectively.

It is important that Vilon and Epitalon modulated the expression of several genes associated with oncogenesis. Vilon and Epitalon inhibited the expression of the myeloblastic oncogene-like gene 1 and the ptotooncogene Bcl-3, respectively.

Table 1. Effects of Vilon and Epitalon on gene expression in female CBA mice*

No.	ΔE		Clone	Type	Gene	Category
	Vilon	Epitalon				
Cell division						
1	3.28	3.14	H3056F10	Hs	epidermal growth factor receptor pathway substrate 15 (EPS15)	1a
2	−2.03		H3064C09	Mm	myeloblastic oncogene-like gene 1 (Mybl1)	1a
3		2.50	H3108D12	Mm	gene deleted in polyposis 1 (Dp1)	1a
4		−2.21	H3082H10	Hs	protooncogene Bcl-3 (BCL3)	1a
5	2.12	2.31	H3051A09	Hs	MCM10 homologue	1b
6		−2.11	H3067E06	Mm	DNA primase, p58 subunit (Prim2)	1b
7	3.35	3.08	H3052B05	Hs	culline 5	1d
8	2.34	2.60	H3025B08	Hs	proliferating cell nucleolar antigen P120 (proliferation-associated nucleolar antigen P120, NOL1)	1d
9	2.08	2.24	H3054F09	Mm	cyclin gene ania-6b	1d
10		2.09	H3149G05	Mm	cyclin I (Ccni)	1d
11		−2.06	H3061H06	Mm	activator of S phase kinase (Ask-pending)	1d
12	3.09	2.78	H3052C04	Hs	APG5L, *S. cerevisiae* APG5 homologue (autophagy 5)	1e
13	2.03	2.06	H3054D02	Mm	histone H3, family 3B (H3f3b)	1e
14		−2.14	H3103D04	Mm	gene 1b homologue for *S. cerevisiae* budding uninhibited by benzimidazoles (Bub1b)	1e
Cell signaling systems and communications						
15	−2.79		H3093A07	Mm	platelet ligand selectin (p-selectin, Selpl)	2a
16	2.76	2.66	H3052F06	Mm	platelet/endothelial cell adhesion molecule (Pecam)	2a
17	2.31	2.11	H3054E05	Mm	thrombospondin 3 (Thbs3) and mucin 1 (Muc1)	2a
18	2.52	2.46	H3054H04	Mm	Ca-activated medium/minor K-conducting channel, member 4 of subfamily N (Kcnn4)	2b
19		−2.71	H3097C12	Mm	intracellular mitochondrial Cl channel 4 (Clic4)	2b
20		−2.32	H3100C10	Rn	Ca-activated medium/minor K-conducting channel, member 2 of subfamily N (Kcnn2)	2b
21		−2.19	H3098H10	Mm	calbindin (PCD-29)	2c
22		2.15	H3019D01	Mm	calmodulin (Cam I)	2c
23	2.90		H3001A02	Mm	secretin (Sct)	2d

Table 1. (continued)

No.	ΔE		Clone	Type	Gene	Category
	Vilon	Epitalon				
24		−2.01	H3101E12	Mm	small GTPase gene (Rab11a)	2e
25	3.49	3.27	H3052B11	Mm	protein kinase 3 with PCTAIRE motive (Pctk3)	2g
26	2.40	2.35	H3056E03	Hs	serine/threonine kinase FUSED	2g
27	2.38	2.41	H3052D01	Mm	doublecortin and Ca/calmodulin protein kinase-like 1 (Dcamkl1)	2g
28	2.38	2.25	H3052C12	Mm	serine/threonine kinase 11 (Stk11)	2g
29		2.57	H3047A06	Mm	protein kinase C-z (Pkcz)	2g
Cell structure and motility						
30	3.28	3.02	H3056C01	Mm	formin 2 (Fmn2)	3c
31	2.69	2.66	H3056C12	Mm	dystonin (Bpag1-n)	3c
32		2.32	H3154A09	Mm	filamin	3c
33		−2.21	H3102A05	Mm	centromeric autoantigen H (Cenph)	3c
34	2.81	2.08	H3121F01	Mm	wingless-related MMTV integration site (Wnt4)	3d
35		2.13	H3010B10	Mm	procollagen, type V, a3 (Col5a3)	3d
36		2.26	H3133C10	Hs	Golgi complex autoantigen (golgin, subfamily a, 1; GOLGA1)	3e
Protective systems of cells and organism						
37	3.27	3.09	H3052D11	Mm	adenylate kinase 2 (Ak2)	4b1
38		2.10	H3023E12	Mm	topoisomerase (DNA) III b (Top3b)	4b2
39	3.35	2.94	H3055D08	Mm	protein COP1 (Cop1)	4b3
40	2.37	2.47	H3056G12	Mm	homologue 2 of *R. norvegicus* enigma gene (Enh2-pending)	4b3
41		2.76	H3082E06	Hs	enigma (LIM-domain protein, ENIGMA)	4b3
42		−2.43	H3032G08	Mm	corticosteroid-binding globulin (Cbg)	4b3
43		2.09	H3023H11	Mm	ferritin L-subunit gene	4b3
44		2.09	H3145H02	Hs	translocation protein (TLOC1)	4b3
45	2.49		H3042G07	Mm	heat shock protein 84 (HSP84)	4b4
46		2.42	H3139E01	Mm	70-kDa heat shock protein (HSC70 and HSP73)	4b4
47		2.33	H3054D05	Rn	protein associated with small stress protein PASS1 (Pass1)	4b4
48	3.05	2.96	H3054E04	Mm	class II major histocompatibility protein, chains a and b	4c
49	2.38		H3147A04	Mm	butyrophylin precursor (BT, BUTY)	4c
50	2.23	2.26	H3054E02	Hs	chromosome I mRNA similar to BAT2 genes	4c

Table 1. (continued)

No.	ΔE		Clone	Type	Gene	Category
	Vilon	Epitalon				
51	2.16	2.15	H3055B08	Mm	class III major histocompatibility protein	4c
52		−2.35	H3103C12	Mm	signal lymphocyte activation molecule (Slam)	4c
Expression of genes and proteins						
53	−2.54		H3058C11	Hs	RNA-binding motive protein 9 (RBM9)	5a2
54		−2.26	H3015G02	Mm	RNA-binding motive protein 6 (Rbm6)	5a2
55	3.69	3.27	H3056E09	Hs	zinc finger protein ZNF01 and HUMORFKG1B	5a3
56	3.65	3.19	H3056G09	Mm	Scm-similar gene with 4 mbt-domains (Sfmbt)	5a3
57	3.04		H3133B04	Mm	transcriptional factor 20 (Tcf20)	5a3
58	2.71	2.69	H3054G09	Mm	SMAR1	5a3
59	2.49	3.37	H3133A12	Mm	zinc finger protein 61 (Zfp61)	5a3
60	2.33	3.30	H3116H10	Mm	ets-related transcriptional factor (Etv5)	5a3
61	2.04	2.38	H3139D10	Hs	thyroid hormone receptor-interacting protein 12 (TRIP12)	5a3
62	2.01		H3098G05	Mm	DNA-binding protein inhibitor ID-2	5a3
63		2.46	H3131F02	Hs	transcription-enhancing MADS-box factor 2, polypeptide B (myocyte-enhancing factor 2B, MEF2B)	5a3
64		−2.05	H3074C11	Hs	TFIIB-related factor 2	5a3
65	3.08	2.99	H3047B05	Hs	FLJ12848 fis, highly homologous to Hs mRNA for nuclear transport receptor	5b1
66		3.03	H3080D01	Mm	polyubiquitin C (Ubc)	5b1
67		−2.13	H3103H06	Mm	neighbor of A kinase-anchoring protein 95 (Nakap95-pending)	5b1
68		−2.07	H3068G02	Mm	ubiquitin-conjugating enzyme 2e (Ubc2e)	5b1
69	−2.47	−2.16	H3049A01	Hs	SUMO-1-specific protease FKSG6	5b2
70	−2.06		H3042F12	Mm	serine protease inhibitor 4 (Spi4)	5b2
71	2.06	2.75	H3047B07	Mm	tripeptidyl peptidase II (Tpp2)	5b2
72	2.94	2.88	H3054F04	Mm	p70/p85 s6 kinase	5b3
73		2.46	H3011E01	Mm	ribosomal protein L8 (Rpl8)	5b3
74		2.09	H3133B09	Mm	ribosomal protein S6 (Rps6)	5b3
75		2.09	H3126B04	Hs	ribosomal protein S10	5b3
76		2.07	H3126D05	Mm	ribosomal protein L27 (Rpl27)	5b3

Table 1. (continued)

No.	ΔE		Clone	Type	Gene	Category
	Vilon	Epitalon				
Metabolism						
77		2.12	H3148E08	Mm	ornithine decarboxylase antizyme	6b
78		2.26	H3055F07	Mm	*d*-aminolevulinate dehydratase (Lv, ALAD)	6c
79	−2.31	−2.26	H3061H04	Mm	homologue of ATP synthetase b-chain	6d
80	−2.09		H3065E03	Mm	soluble isocitrate dehydrogenase 1 $NADP^+$ (Idh1)	6d
81		2.95	H3005E04	Hs	ATP synthetase b-subunit (ATPSB)	6d
82		2.57	H3001A07	Bt	15-kDa subunit of NADH ubiquinone oxidoreductase, 15 kDa (NIPM)	6d
83		2.50	H3009D02	Rn	ATPase F1F0 d-subunit	6d
84		2.06	H3020D10	Hs	isocitrate dehydrogenase 3 (NAD^+) a (IDH3A)	6d
85	2.91	2.70	H3055C05	Hs	inositol-1,3,4-triphosphate 5/6 kinase (ITPK1)	6e
86	2.17	2.39	H3047D05	Hs	enoyl-CoA hydratase	6e
87		2.15	H3095C08	Mm	dodecenoyl-CoA *d*-isomerase	6e
88	3.18	2.90	H3056E08	Mm	H-type AMP deaminase	6f
89		5.95	H3147A06	Mm	APEX nuclease	6f
90		2.47	H3147B06	Mm	glyceraldehyde-3-phosphate dehydrogenase (Gapd)	6h
91		2.04	H3114C12	Mm	class 3 cytoplasmic aldehyde dehydrogenase (Adh4)	6h
92		2.04	H3145A07	Mm	aldehyde reductase (Akr1A4)	6h
93	2.95	2.73	H3053H06	Mm	citrine (Slc25a13)	6i
94	2.78	2.72	H3052B06	Mm	ATP-binding cassette, subfamily B (MDR/TAP), member 1 (Abcb1)	6i
95	2.42	2.54	H3033B04	Hs	family of soluble carriers 7 (cationic amino acid transporter, system y+), member 6 (SLC7A6)	6i
Unclassified genes						
96	5.34	5.64	H3047A02	Mm	serine/threonine protein kinase of protooncogene A-Raf (KRAA)	7a
97	4.91	5.18	H3047H01	Mm	Nulp1	7a
98	3.38	3.07	H3056D02	Hs	FLJ22439 fis	7a
99	3.20	2.85	H3056G05	Hs	protein KIAA0029	7a
100	3.19	3.06	H3052B09	Mm	nuclear antigen Sp100	7a
101	3.12	2.97	H3055D05	Hs	protein KIAA0970	7a

Table 1. (continued)

No.	ΔE		Clone	Type	Gene	Category
	Vilon	Epitalon				
102	3.05	2.91	H3056E07	Hs	FLJ13697 fis	7a
103	2.93	2.78	H3052C02	Hs	KIAA0308 gene	7a
104	−2.77		H3095C07	Mm	channel-interacting PDZ-domain protein (Cipp)	7a
105	2.67	2.60	H3053G06	Mm	mg53d08.r1	7a
106	2.61	2.75	H3039H10	Mm	epithelial protein lost in neoplasm (Eplin)	7a
107	2.60	2.29	H3056E11	Hs	brefeldin A-inhibited guanine nucleotide-exchange protein 2 (BIG2)	7a
108	2.59	2.81	H3054E11	Hs	phosphoinositol-3-phosphate-binding protein 2 (PEPP2)	7a
109	2.55	2.53	H3052C05	Hs	putative sialoglycoprotease type 2 (LOC64172)	7a
110	−2.29		H3015A03	Hs	protein CGI-63 (LOC51102)	7a
111	2.28	3.07	H3047F05	Hs	gene 3 for translocational myeloid/ lymphoid or mixed-lineage leukemia, trithorax (*Drosophila*) homologue, MLLT3	7a
112	2.28	2.41	H3056H01	Hs	MSTP028	7a
113	2.25	2.48	H3052G11	Mm	ganglioside-induced differentiation-associated protein 1 (Gdap1)	7a
114	2.21	2.20	H3054H11	Hs	FLJ10977 fis	7a
115	2.14		H3033B01	Hs	FLJ22386 fis	7a
116	2.13	3.64	H3047C01	Hs	hypothetical protein FLJ10914	7a
117	2.13	2.14	H3053G04	Hs	hypothetical protein FLJ20476	7a
118	−2.12		H3002A11	Hs	hypothetical protein FLJ10252	7a
119	2.11	2.50	H3002D05	Hs	FLJ12166 fis	7a
120	2.07	2.28	H3054C10	Hs	protein KIAA0699	7a
121		2.94	H3074A01	Hs	KIAA0182	7a
122		2.83	H3021G11	Mm	calreticulin (Calr)	7a
123		−2.62	H3142H11	Hs	FLJ22230 fis	7a
124		−2.56	H3097C06	Mm	transforming growth factor-β 1-induced transcript 4 (Tgfb1i4)	7a
125		−2.49	H3102C10	Mm	ERIC1 (Eric1)	7a
126		2.48	H3158F09	Hs	FLJ22903 fis	7a
127		2.37	H3084E12	Hs	FLJ21480 fis	7a
128		2.30	H3139C11	Mm	MS4A11	7a
129		2.19	H3022D10	Hs	brain acid-soluble protein 1 (BASP1)	7a
130		2.18	H3144C06	Mm	24.6-kDa protein	7a

Table 1. (continued)

No.	ΔE		Clone	Type	Gene	Category
	Vilon	Epitalon				
131		2.16	H3007C12	Hs	hypothetical protein HSA011916	7a
132		2.15	H3095C05	Mm	Shfdg1	7a
133		−2.14	H3060A02	Mm	cysteine-rich repeat-including protein CRIM1 (Crim1)	7a
134		−2.12	H3061F04	Hs	protein HYA22	7a
135		2.08	H3018F08	Hs	hypothetical protein FLJ20419	7a
136		−2.07	H3063C10	Mf	brain clone QnpA-21065	7a
137		−2.07	H3097G04	Mm	protein MLN 64 (protein ES 64, ML64)	7a
138		−2.06	H3097G02	Hs	protein KIAA1157	7a
139		2.03	H3136B04	Mm	uterine protein LOC55978	7a
Mitochondrial genes						
140	6.13	6.61	H3139C10	Mm	mitochondrial gene 16S	Mt
141	−2.25		H3049A03	Mm	ATPase 6 mitochondrial gene	Mt
142		2.32	H3024A09	Mm	NADH dehydrogenase 5 mitochondrial gene (ND5)	Mt
143		2.19	H3023G11	Mm	cytochrome b mitochondrial gene (CYT b)	Mt
144		2.03	H3011D05	Mm	NADH dehydrogenase 4 mitochondrial gene (ND4)	Mt

*Table includes only identified genes.
ΔE = Changes in gene expression; Hs = *Homo sapiens*; Mm = *Mus musculus*; Rn = *Rattus norvegicus*; Mf = *Macaca fascicularis*; Bt = *Bos taurus*.

These peptides activated the expression of genes for protein kinase C-zeta (Epitalon), LIM/PDZ-domain enigma proteins (Epitalon), and homologue 2 (both peptides), which are probably involved in oncogenesis and suppress the expression of functionally related genes for PDZ-domain Cipp protein (Vilon) [Kurschner et al., 1998; Bach, 2000; Schondorf et al., 2001]. It should be emphasized that these peptides affect genes that are related to calcium exchange. Both peptides increase the expression of culline-5 and the genes for Kcnn4 and Dcamkl1. Epitalon activates the expression of calmodulin, but inhibits the expression of genes for Ca-binding protein calbindin and Kcnn2. Vilon and Epitalon increase the expression of genes for serine/threonine kinases Pctk3, FUSED, and Stk11 which belong to the same functional category (cell signaling systems and communication). At least one of these kinases

(Stk11) with unknown functions possesses anticarcinogenic activity. Mutations in its gene lead to the development of Peits-Gingers syndrome, which increases the risk of tumors [Hemminiki, 1999]. These data are consistent with the inhibitory effects of Epitalon and Vilon on the development of spontaneous tumors.

Thus, specific effects were revealed of the peptide bioregulators Epitalon and Vilon on gene expression. At the same time, it is evident that in experiments with the whole organism in vivo it would be difficult to determine whether the effect of the particular factor on a particular organ is direct or if it is caused by the effect on some other tissues, and it is they which affect the studied organ. Moreover, the revealed changes in the expression of genes can be caused by hormonal background alterations or cytokines secretion.

3

Effect of Peptides on IL-2 Gene Expression

3.1. Effect of Vilon and Epitalon on IL-2 Gene Expression in Mouse Spleen Lymphocytes

The effects of low-molecular-weight immunomodulators on the expression of cytokine genes are poorly studied.

It is known that the functional activity of immunocompetent cells decreases under the effect of extreme factors and during aging [Korneva and Shkhinek, 1988]. It was shown that IL-2 gene expression in lymphocytes, which yields the protein playing a key role in immune functions and T lymphocyte response to mitogens, decreases during aging [Levine and Prystowsky, 1995]. Studies of IL-2 gene expression allow one to reveal metabolic disturbances at the early stages of cell activation and help to understand the mechanisms of the immunomodulating effects of peptide bioregulators. Various peptides from the thymus and pineal gland, myelopeptides, and opiate peptides affect the immune response, phagocytic and antigen-presenting functions of macrophages, and the production of IL-1, IL-2, IL-6 and tumor necrosis factor-α [Chipens et al., 1987; Cheido et al., 1993; Morozov and Khavinson, 1996]. N-Acetylmuramyl-*L*-Ala-*D*-isoGln (muramyl dipeptide) and its derivatives stimulate gene expression and synthesis of IL-1α, IL-1β, IL-6, IL-8 and TNF in blood mononuclear cells [Darcissac et al., 1996; Worth et al., 1999]. Trans-acting factors of IL-2: NF-AT, NF-kB, AP-1, AP-2, AP-3, Oct-1 and others, consisting of high-molecular-weight components were extensively studied [Garrity et al., 1994]. It was shown that analogues of hypothalamic peptide hormone luliberin, Ns-Pro-Ser-Tyr-D-Asp-Leu-Arg-Pro-NHEt

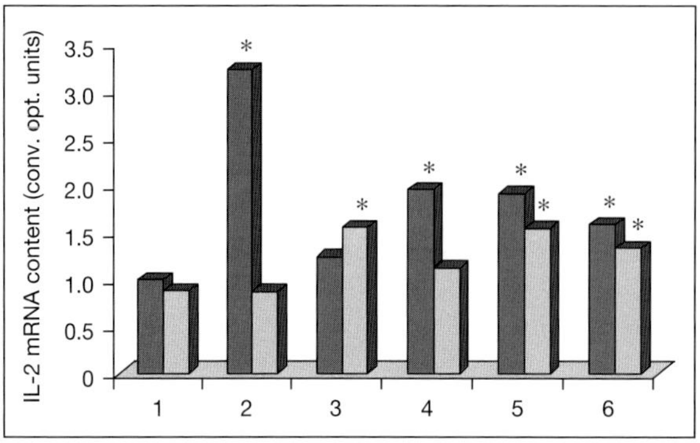

Fig. 2. Effect of Vilon on IL-2 mRNA synthesis in mouse spleen lymphocytes. After 5 h (■) and 20 h (□) of incubation. 1 = Control; 2 = Con A + rIL-2; 3 = Vilon (50 pg/ml/million cells); 4 = Vilon (5 ng/ml/million cells); 5 = Vilon (50 ng/ml/million cells); 6 = Vilon (100 ng/ml/million cells). $^*p < 0.05$ as compared to the corresponding index in the control.

and H-Pro-D-Phe-Pro-Ser-Tyr-D-Lys-Leu-Arg-Pro-Gly-NH_2 stimulate IL-2 mRNA synthesis and differentiation of mouse spleen T cells [Kazakova et al., 1996]. Low-molecular-weight peptides with the same N-terminal sequence Pro-Glu-Pro-Ala-Lys-Ser-Ala-Pro-Ala-Pro, isolated from the nuclei of spleen and brain cells of immunized animals have similar properties [Grishina et al., 1996].

We studied the effects of peptides Vilon (Lys-Glu) and Epitalon (Ala-Glu-Asp-Gly) on IL-2 gene expression in mouse spleen lymphocytes in vitro. The effect of the peptides was studied in 50 pg/ml, 5, 50 and 100 ng/ml concentrations. Unstimulated lymphocytes and cells incubated with concanavalin A (Con A, 5 mg/ml) and rIL-2 (30 U/ml) served as the control. During the study, pAA1213 plasmid containing a full-length cDNA copy of the human IL-2 gene was used (550 nucleotide sequences). IL-2 mRNA was assayed in spleen lymphocytes from seventeen 10- to 12-week-old male CBA mice weighing 14–16 g. Lymphocytes were isolated and incubated at 37°C for 5 or 20 h. The time limits for incubation were chosen according to experimental data, which show that the expression of IL-2 mRNA in T lymphocytes after the effect of inductors both in in vitro conditions and in vivo, reaches its maximum at 8 h and stops at 24 h [Mouzaki et al., 1991; Rogers et al., 1997].

It was revealed that after 5 h incubation Vilon and Epitalon stimulated IL-2 mRNA synthesis in lymphoid cells (fig. 2, 3), especially at the concentrations of 5 ng/ml (Vilon) and 50 pg/ml (Epitalon). Prolonged (20 h) incubation produced less-pronounced changes in IL-2 gene expression, but this parameter remained above the control level.

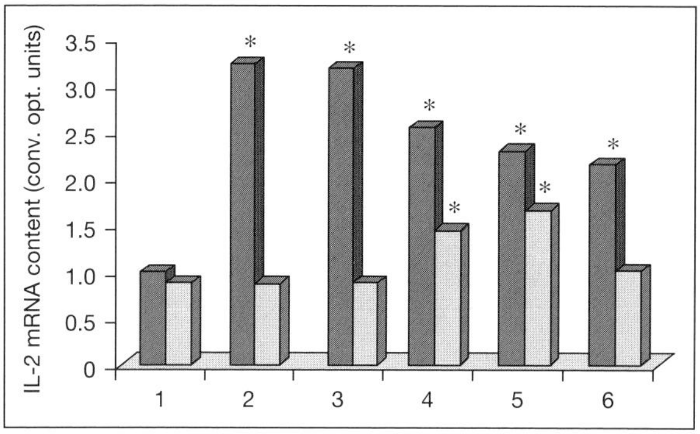

Fig. 3. Effect of Epitalon on IL-2 mRNA synthesis in mouse spleen lymphocytes. After 5 h (■) and 20 h (□) of incubation. 1 = Control; 2 = Con A + rIL-2; 3 = Epitalon (50 pg/ml/mln cells); 4 = Epitalon (5 ng/ml/mln cells); 5 = Epitalon (50 ng/ml/mln cells); 6 = Epitalon (100 ng/ml/mln cells). * $p < 0.05$ as compared to the corresponding index in the control.

Thus, both peptides, Vilon and Epitalon, stimulate the in vitro synthesis of IL-2 mRNA in mouse spleen lymphocytes. This effect depends on the peptide concentration and the duration of incubation. In addition, these peptides activate IL-2 gene expression in cultured lymphocytes without additional stimulation with Con A and rIL-2. It should be emphasized that IL-2 gene expression in cells increased after 20 h incubation with the peptides, when the addition of mitogens into the culture medium had no stimulating effect. Probably, the presence of these peptides in the incubation medium prolonged the effects of nuclear transacting factors regulating IL-2 gene expression.

It is known that transcription regulation in eukaryotes is a complex process including many stages of positive and negative regulation of gene expression. These processes include translocation of a signal from the cell surface to the cytoplasm and its transduction to nuclear genetic structures. The stimulatory effect of Vilon and Epitalon on IL-2 mRNA synthesis in mouse splenocytes not only confirms the possibility of transporting dipeptides across the cell membrane, but also indicates their involvement in the regulation of gene expression.

These data suggest that peptides Vilon (Lys-Glu) and Epitalon (Ala-Glu-Asp-Gly) are the shortest regulatory fragments formed in cells during organic proteolysis and performing specific regulatory functions: they are involved in the transport of transacting factors into the nucleus or enter the composition of their active centers necessary for the activation of IL-2 gene transcription in lymphocytes.

3.2. Effect of Vilon and Epitalon on IL-2 Gene Expression in Rat Hypothalamus Structures

IL-2 is a protein product, which plays a key role in the development of the immune system functions. It is synthesized in nervous system cells [Sei et al., 1995; Campbell et al., 1998].

It is known that hypothalamic structures participate in the regulation of different vegetative and immunologic functions of the organism. During electrostimulation of the lateral hypothalamic field (LHA) there was a registered increase in natural killer cell activity [Wenner et al., 2000]. The ventromedial nucleus of the hypothalamus (VMH) is an LHA antagonist, which exerts an inhibitory effect on natural killer cell activity. It was revealed that destruction of the posterior hypothalamus structures affects the intensity of the immune response [Korneva, 1969].

This assay was aimed at studying the effect of Vilon (Lys-Glu) and Epitalon (Ala-Glu-Asp-Gly) on IL-2 gene expression intensity in structures of rat brain hypothalamus. The experiments were conducted on 27 Sprague-Dawley male rats weighing 180–200 g. The preparations were administered daily intramuscularly at a dose of 10 μg/kg during 5 days.

During the study, we used a plasmid-containing full-size cDNA replica of the human IL-2 gene. IL-2 mRNA analysis was conducted in structures of rat brain hypothalamus by in situ hybridization on paraffin sections of brain after intracardial perfusion. Animals were anesthetized with phenobarbital (60 μg/kg in 1 ml of physiologic salt solution) and intracardial perfusion was conducted using a warm physiologic solution (100 μl) containing 20 U/ml of heparin, with 10–15 ml/min flow speed, and then with fixating solution (250 ml) (4% parapharmaldegid in 0.1 *M* PBS and 0.2% picrine acid, pH 7.4). Brain isolation was conducted 1 h after perfusion. Paraffin brain tissue sections were used to estimate the results.

The results of the study show that only under the effect of Epitalon (when these particular methods and terms of administration are used) is activation of the neuronal cells registered. After Epitalon administration in structures of the hypothalamus such as the lateral hypothalamic field, the front hypothalamic field, the dorsomedial nucleus, the paraventricular nucleus and the hypothalamic amygdaloid complex, cells containing mRNA IL-2 were registered (fig. 4–8).

During Vilon administration in the studied structures single cells were revealed containing mRNA IL-2. The introduction of physiologic solution as a control substance did not result in the induction of mRNA IL-2 synthesis in these particular structures.

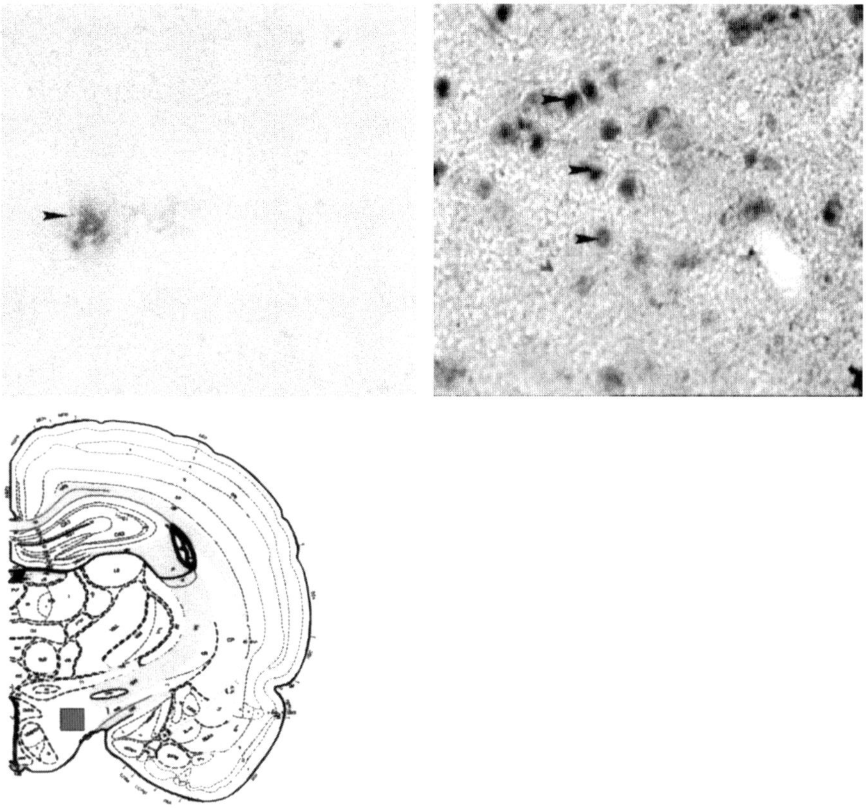

a *b*

Fig. 4. Effect of Vilon (***a***) and Epitalon (***b***) on IL-2 mRNA synthesis in the lateral area cells of rat hypothalamus. ×40.

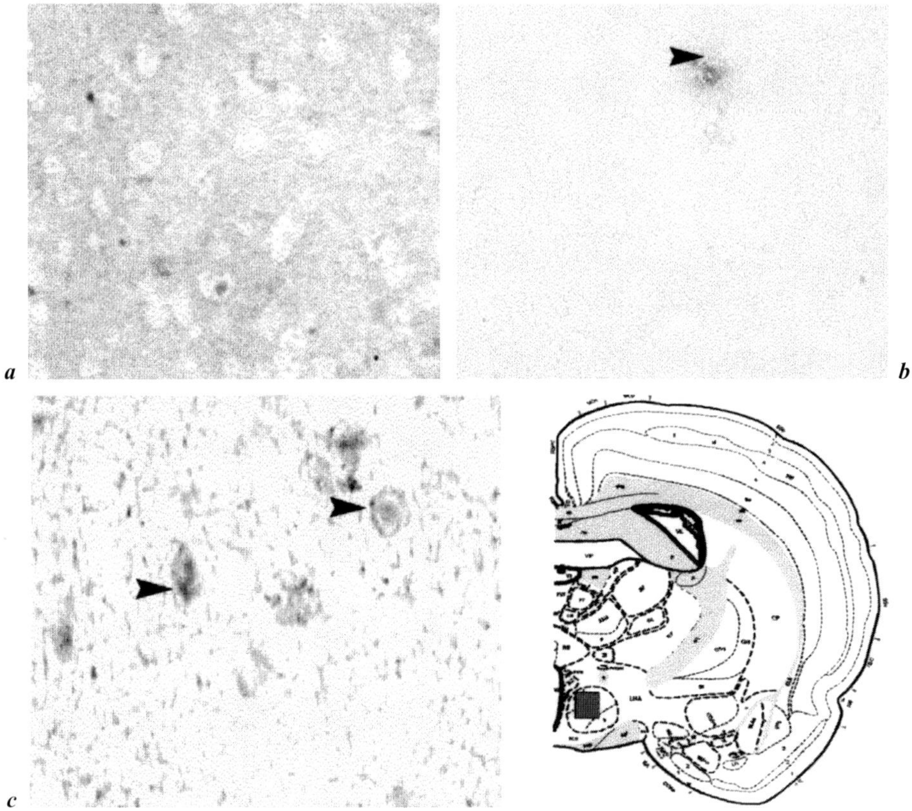

Fig. 5. Effect of Vilon (***b***) and Epitalon (***c***) on IL-2 mRNA synthesis in anterior area cells of rat hypothalamus. (***a***) = Control. ×40.

a

b

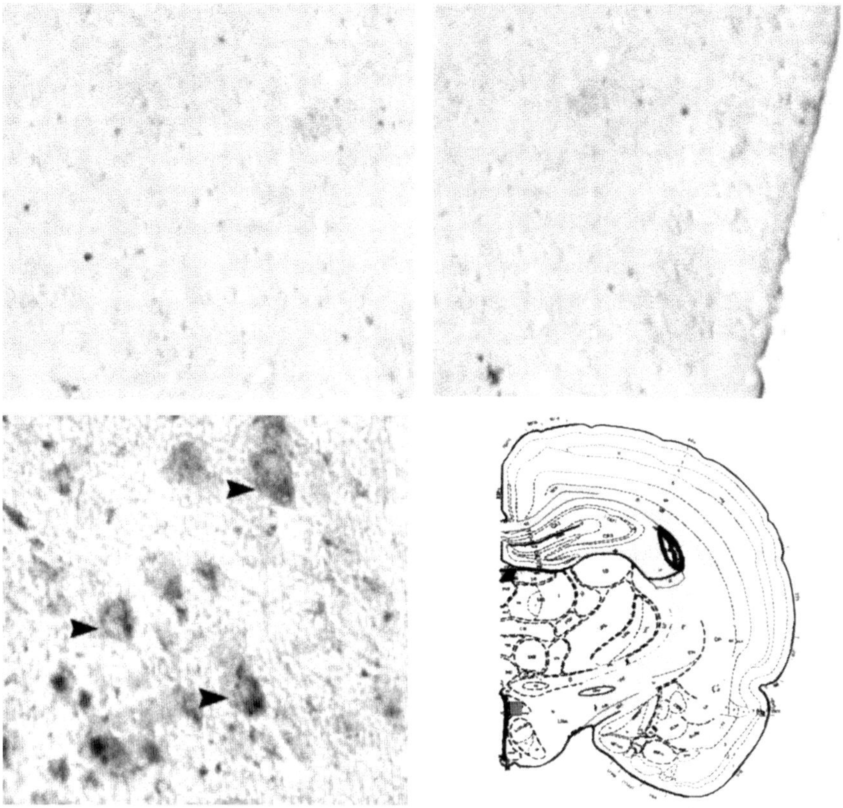

c

Fig. 6. Effect of Vilon (***b***) and Epitalon (***c***) on IL-2 mRNA synthesis in dorso-medial cells of rat hypothalamus. (***a***) Control. ×40.

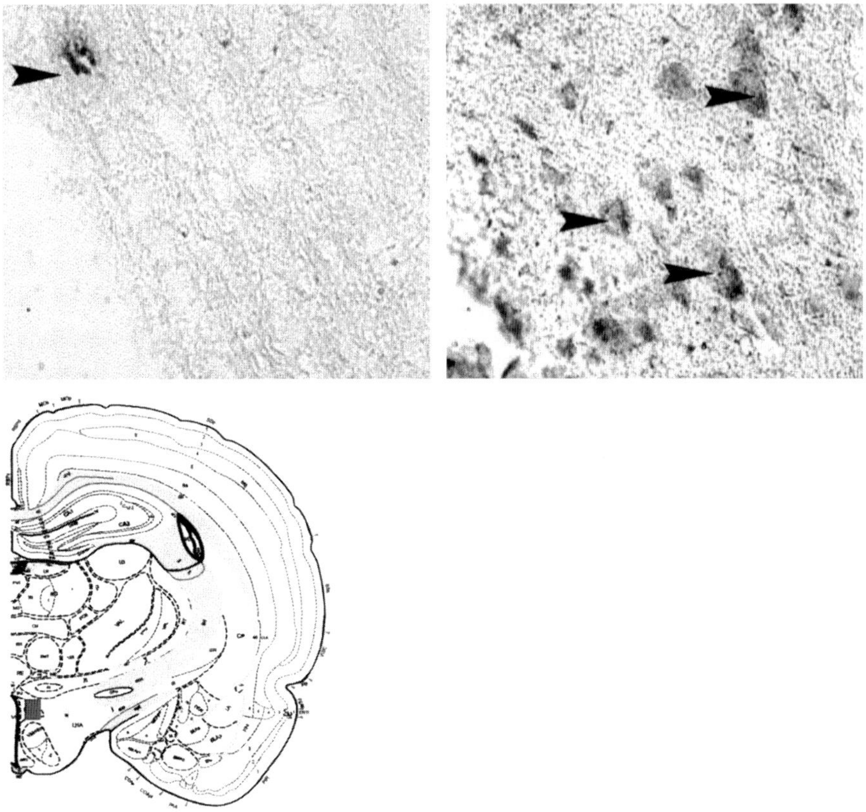

a *b*

Fig. 7. Effect of Vilon (***a***) and (***b***) Epitalon on IL-2 mRNA synthesis in paraventricular nucleus cells of rat hypothalamus. ×40.

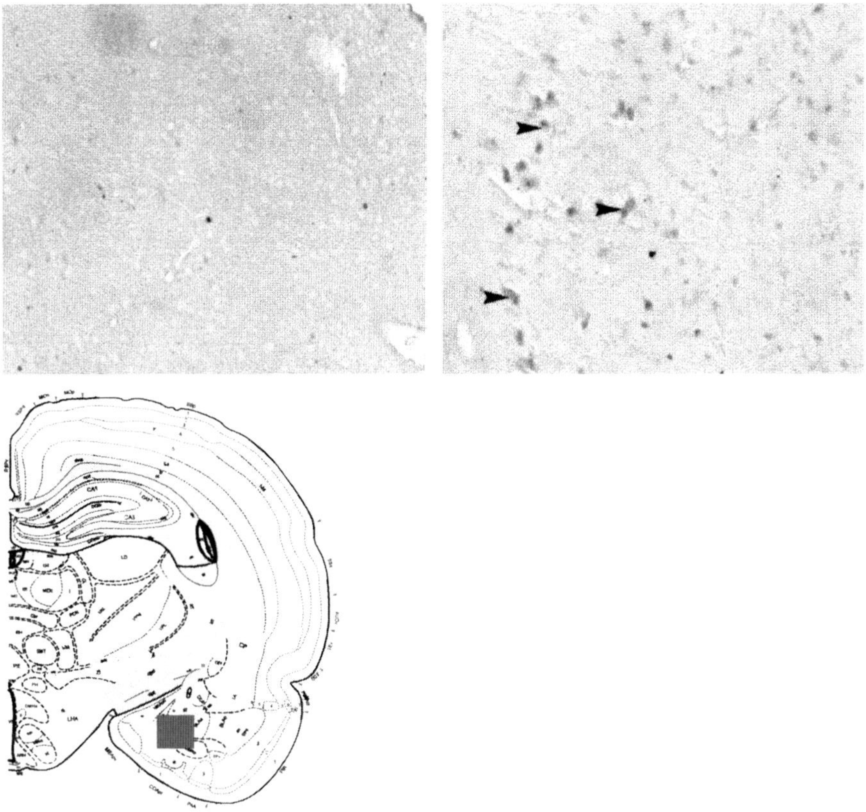

a *b*

Fig. 8. Effect of Epitalon on IL-2 mRNA synthesis in amigdaloid complex cells of rat hypothalamus. (***a***) Control. (***b***) Epitalon. ×10.

4

Effect of Peptides on *c-fos* Gene Expression in Stress

4.1. Effect of Vilon on *c-fos* Gene Expression in Neurons of Rat Hypothalamic Paraventricular Nucleus under Emotional Stress

Emotional stress is known to intensify early genes expression in different brain structures [Anochin and Ryabinin, 1993; Melia et al., 1994; Senba et al., 1994]. An accepted marker of neural cells activation is the c-Fos protein, a product of the immediate early response gene, which functions as a mediator between the process of transcription change under the effect of stimulus and the process of informational signal realization into long-term phenotypical changes [Bullitt, 1990; Hunt et al., 1987; Oladehin, Blatteis, 1995; Yokoyama and Sasaki, 1999]. *Fos* proteins become active at different times after exposure to stimulus. Thus, mRNA *c-fos* expression already starts in several minutes and reaches its maximum in 30–60 min. The maximum synthesis of the *c-fos* protein is observed between the 1st and the 3rd hour and 4–6 h after exposure to the stimulus. In neurons, a decrease of protein synthesis takes place. The *c-fos* gene expression in the brain of stress-resistant and stress-prone animals is different [Sudakov, 1998]. The most intensive *c-fos* gene expression in brain structures (cortex, tonsil and truncus cerebri) was revealed in rats predisposed to emotional stress. In stress-resistant rats under the same experimental conditions, *c-fos* gene expression was registered only in the infralimbal cortex and the olfactory nuclei.

Based on the above results, the aim of the experiments was to study the effect of Vilon (Lys-Glu) on *c-fos* gene expression in the paraventicular nucleus of hypothalamus in stress-resistant and stress-prone animals, and in rats exposed to emotional stress.

The experiments were conducted on Wistar rats. The behavior of the animals was estimated in the open field in order to make a prognosis of the level of their resistance to emotional stress. The effectiveness of Vilon was studied in stress-resistant and stress-prone rats. Vilon was administered intraperitonealy at the doses of 0.1 and 0.01 μg per rat 7–10 days after the first test in the open field. One hour after the injection, the rats were exposed to 60 min of emotional stress, which was modulated by motional restriction.

To identify the cells containing Fos protein, a product of the early *c-fos* gene expression, an immunohystochemical investigation was performed on the brain section with automated counting of the Fos-positive cells. The peculiarities of *c-fos* gene induction were studied in the hypothalamic paraventricular nucleus.

These studies revealed that the number of Fos-positive cells in the hypothalamic paraventricular nucleus in rats which were considered to be stress-resistant was pronouncedly higher than in the animals which were predisposed to emotional stress. Both groups (exposed to and not exposed to the stress) had the same results.

Vilon administration at the dose of 0.01 μg led to an increase in Fos-positive neural cells number in the stress-resistant rats which were not exposed to the stress and in those which were exposed to the stress. At the same time, Vilon at a dose of 0.1 μg led to an increase in Fos-positive neural cells number in the hypothalamic paraventricular nucleus both in the stress-resistant and the stress-prone rats. Against a stress background, Vilon at a dose of 0.1 μg increased the number of Fos-positive neurons in the stress-resistant animals.

Thus, the dose-dependent stress protective effects of Vilon were revealed, increasing *c-fos* gene expression and activating neurons and forming adaptation reactions.

4.2. Effect of Vilon and Epitalon on *c-fos* Gene Expression in Cells of Different Hypothalamic Structures in Rats under 'Handling' Stress

In order to determine the effects of Vilon (Lys-Glu) and Epitalon (Ala-Glu-Asp-Gly) on the intensity of *c-fos* gene expression in cells from different brain structures of rats under stress, we used different methods of peptide administration.

The animals were subjected to single intramuscular injections (2 μg per rat) or single intranasal administration (1 μl (10 ng/μl) in turn into each nasal cavity). The animals of the control group were administered physiologic salt solution. The peptide effects were analyzed 24 h after intramuscular injections and at 30 min and 2 h after intranasal administration.

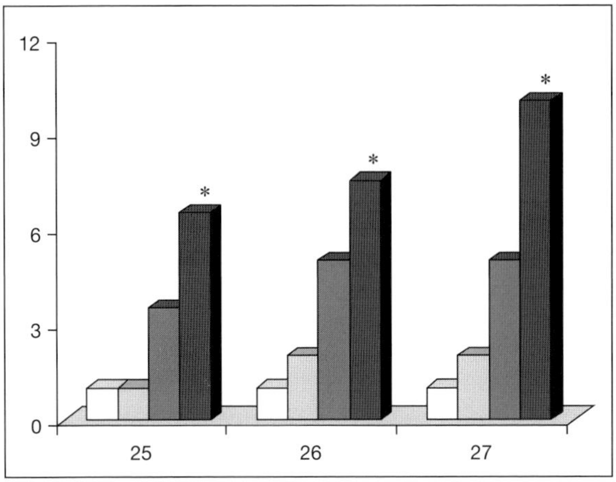

Fig. 9. Quantity of cells containing c-Fos-similar protein in LHA of stress-exposed animals 30 min after intranasal infusion of Vilon and Epitalon. Ordinate shows the quantity of c-Fos in immunoreactive cells in the studied structure of hypothalamus, and the abscissas gives the 25–27th levels of brain sections according to the stereotaxic atlas of rat brain maps [Swanson, 1992]. Intact animals (□); control (NaCl solution) (□); Vilon (■); Epitalon (■). $^*p < 0.05$ as compared to the index in the control.

To modulate stress conditions, rats were handled softly for 3 min ('handling'). Intact animals and those which underwent a 5-day adaptation to the researcher's hands, as well as the non-adapted and adapted rats administered physiologic salt solution, served as a control group. 'Handling' (taking the animal out of the cage, handling, fixing in hands) is a pronounced stress factor, which leads to the activation of brain neurons and *c-fos* gene expression [Kovacs, 1998].

c-fos gene expression was analyzed in different hypothalamic cells of the animals of each group. We studied the following sections of the brain: hypothalamic lateral field (LHA), anterior (AHN), paraventricular (PVH), ventramedial (VMH), dorsomedial (DMH), supraoptical (SO), suprachiasmatic (SCH), arcuate (ARH) hypothalamic nuclei, and the medial amygdal nucleus (MEA). c-Fos-like protein was revealed by an indirect immunoperoxidase method with polyclonal antibodies against the family of c-Fos proteins (Santa-Cruz, Biotech.) (diluted 1:200). Antirabbit IgG labelled with peroxidase (Sigma) diluted 1:300 was used as the secondary antibodies. The number of c-Fos-immunoreactive cells was estimated on the area of the analyzed brain structure (10,000 μm^2).

Figures 9 and 10 demonstrate the results of *c-fos* gene expression in LHA (fig. 9) and SO, PVH, VMH (fig. 10) 30 min after intranasal administration of the peptides. The analysis of LHA consecutive sections revealed stimulation of c-Fos-like protein synthesis mostly after the administration of Epitalon. Vilon

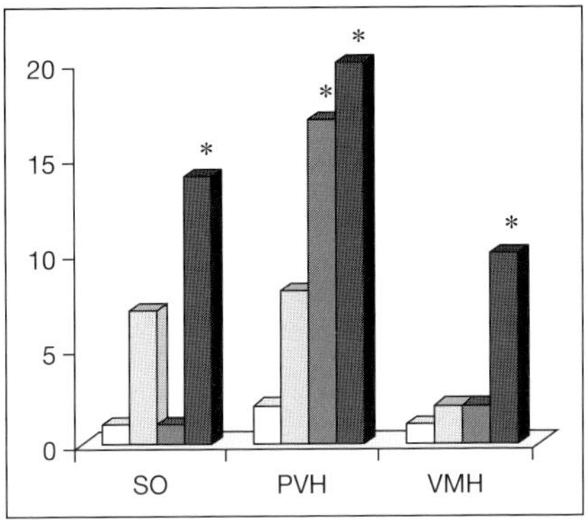

Fig. 10. Quantity of cells containing c-Fos-similar protein in cells of different hypothalamus structures in stress-exposed animals 30 min after the intranasal infusion of Vilon and Epitalon. The quantity of c-Fos in immunoreactive cells in studied structure of hypothalamus is given on the ordinate, and the 24th level of brain sections according to the stereotaxic atlas of rat brain maps on the abscissa [Swanson, 1992]. Intact animals (□); control (NaCl solution) (□); Vilon (■); Epitalon (■). $^*p < 0.05$ as compared to the control.

stimulated c-Fos-like protein synthesis in PVH cells. Epitalon positively affected *c-fos* gene expression 30 min after administration to the SO, PVH and VMH cells.

Analysis of the c-Fos-like protein synthesis in cells of different hypothalamic structures and MEA 2 h after intranasal administration of peptides revealed the most pronounced stimulation under the effect of Epitalon in the AHN, SO, SCH and MEA cells, while Vilon significantly decreased the number of c-Fos-positive cells in all the studied structures, except the SCH and MEA (fig. 11).

Therefore, the results of this study demonstrate a stimulating effect of the 'handling' method in combination with the intranasal administration of physiologic salt solution on *c-fos* gene expression 2 h after administration in all the analyzed hypothalamic structures. Intranasal administration of the peptides to the rats, subjected to the 'handling', mostly led to different results: Vilon application at 2 h decreased the *c-fos* expression in the AHN, PVH and LHA. Epitalon administration both at 30 min and in 2 h increased c-Fos-like protein synthesis in the AHN, SO, SCH, VMH and MEA as compared to the reaction on physiologic salt solution administration combined with 'handling'. It should be noted that Epitalon as well as Vilon inhibited the increase of *c-fos* expression (caused by stress) in the neurosecretory hypothalamic nucleus (PVH). A pronounced c-Fos-like protein synthesis stimulation in LHA and PVH was registered during a short period of Epitalon administration (30 min) contrary to the long period (2 h).

During the next series of experiments, we studied *c-fos* gene expression in cells of different hypothalamic structures of rats 24 h after 'handling' and single

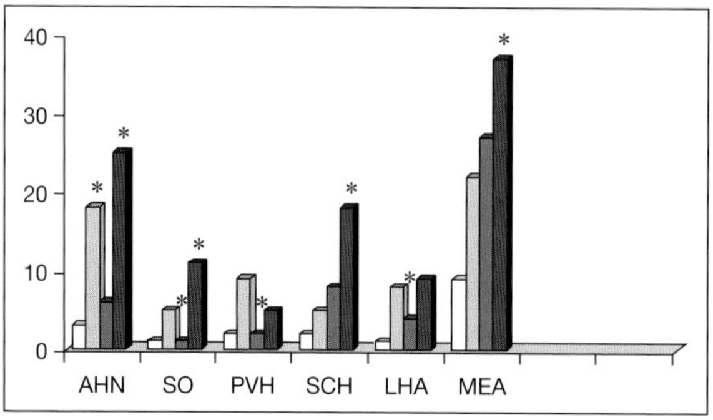

Fig. 11. Quantity of cells containing c-Fos-similar protein in cells of different hypothalamic structures in stress-exposed animals 2 h after intranasal infusion of Vilon and Epitalon. Quantity of c-Fos in immune reactive cells in the studied structure of the hypothalamusis given on the ordinate, and the 24th level of brain sections according to the stereotaxic atlas of rat brain maps on the abscissa [Swanson, 1992]. Intact animals (□); control (NaCl solution) (□); Vilon (■); Epitalon (■). *$p < 0.05$ as compared to the control.

intramuscular injection of Vilon or Epitalon. The experiment was conducted on animals which were non-adapted to experimental conditions and subjected to 'handling'. As can be seen from table 2, a single injection of physiologic salt solution 24 h after 'handling' stress inhibited c-Fos-like protein synthesis stimulation in the majority of the analyzed hypothalamic structures (AHN, SCH, SO, LHA, VMH, PVH, DMH, ARH).

Administration of Vilon under these conditions reduced the number of c-Fos-containing cells in the AHN and SCH and increased that in the PVH compared to the same indices registered in animals after physiologic salt solution injection. Epitalon increased the number of c-Fos-containing cells in the PVH, SO and ARH compared to the number of activated cells which appear after the injection of physiologic salt solution.

It should be noted that PVH is the only hypothalamic structure, among those analyzed, which showed that peptides only directly affected the *c*-fos gene expression. Single intramuscular administration of Vilon and Epitalon to the rats subjected to 'handling' stimulates synthesis of c-Fos-like protein in PVH cells – hypothalamic nucleus, associated with regulation of organism stress reactions [Honkaniemi et al., 1992]. The previous experiment, which was conducted 2 h after 'handling' and intranasal administration of peptides, revealed a decrease in the number of c-Fos-containing cells in the PVH as compared to the indices of gene activation in the control animals. This difference can be explained by the different time periods after administration and the specificity of PVH functioning. At 2 h, presumably, a direct inhibiting effect of peptides appears on the *c-fos* gene expression, while at 24 h the peptide effect is already indirect.

Table 2. Number of cells expressing c-fos-like protein in different hypothalamic structures of rats 24 h after 'handling' and single Vilon and Epitalon intramuscular injection

Groups of animals	Number of c-fos-positive cells in the structures							
	AHN	SCH	SO	LHA	PVH	VMH	DMH	ARH
	Levels of brain sections							
	24	24	25	25	26	27	28	28
Intact rats	1 ± 0.3	2 ± 0.1	2 ± 0.1	2 ± 0.1	2 ± 0.5	2 ± 0.5	3 ± 0.5	3 ± 1
Control (physiologic salt solution)	15 ± 1*	19 ± 1*	4 ± 0.9	9 ± 0.4*	6 ± 0.2*	2 ± 0.3	15 ± 1*	9 ± 0.6*
Vilon	3 ± 0.9*	0	4 ± 0.1	2 ± 0.2	20 ± 1.1*#	1 ± 0.1	9 ± 0.5*#	5 ± 1.2
Epitalon	25 ± 2*#	15 ± 0.9#	14 ± 1*#	2 ± 0.1	20 ± 0.5*	6 ± 0.3*	12 ± 2*	16 ± 2*#

*$p < 0.05$ – as compared to the intact animals indices; #$p < 0.05$ – as compared to the control indices. Brain sections levels are given according to the stereotaxic rat brain atlas [Swanson, 1992].

As for the cells of other studied hypothalamic structures, only Epitalon stimulates c-Fos protein synthesis in them. Vilon injection, combined with the 'handling' and stimulating effect of the physiologic salt solution, inhibits the synthesis of c-Fos-like proteins in AHN and SCH.

Thus, studies of the *c-fos* immediate early response gene expression in hypothalamic structures under weak stress and the analysis of Epitalon and Vilon effect allowed evaluation of the character of stress-induced disorders in the nervous system cells and a correcting effect of the peptides. A spatial-temporal pattern was determined of the hypothalamic structure activation in case of stress ('handling'), and also in response to non-antigenic stimuli such as intranasal or intramuscular peptide administration. A dependence was defined of the process of c-Fos-like protein expression on the method and time of peptide administration and the level of the section of the analyzed hypothalamic structure.

4.3. Effect of Epitalon on c-Fos Protein Content in the Pineal Gland of Rats Exposed to Osmotic Stress

Both the pineal gland and the hypothalamus-hypophysis complex take part in the formation of the general adaptation syndrome [Anisimov, 1997; Arushanyan et al., 1993] Stress makes the pineal gland to increase the secretion of melatonin and peptide substances [Arushanyan et al., 1993; Kovalenko et al., 1997], which are known to reduce the damaging effect of oxygenic, osmotic, psychic and other stresses. The pineal gland belongs to circumventricular organs which do not have a blood-brain barrier. Consequently, this organ is highly sensitive to macromolecular biologically active substances circulating with the cerebral blood flow and, especially, to peptides. The unique location of the olfactory system, its chemical links both to the environment and the central nervous system turn it into a convenient pathway for the noninvasive delivery of substances to the cerebral blood flow and circumventricular organs. There is a constant liquor diffusion through the subarachnoid space along the olfactory neurons to the nasal submucosa and nasal lymph system [Kida et al., 1993]. This is a possible way for metals, dyes, viruses, peptides, proteins and narcotics to enter the brain via the nasal cavity avoiding the hemato-encephalic barrier [Pietrowsky et al., 1996a,b]. It was previously revealed that intranasally infused epiphyseal peptides reach the pineal gland and specifically regulate its electrical activity and pinealocytic ultrastructure. Furthermore, these effects have been observed only under conditions of stress but not in the norm [Sibarov et al., 2002a, b]. An important role in intracellular pineal synthesis activation, for example in stress, belongs to the heterodimer AP-1 formed out of the *c-fos* and *c-jun* transcription factors [Kovaks, 1998].

To prove the hypothesis on the participation of oncogenes in pineal humoral self-regulation, an immunohistochemical detection of c-Fos protein in the pinealocytes was performed in the norm, in stress-exposed rats and in the case of Epitalon (Ala-Glu-Asp-Gly) intranasal infusion.

To detect the c-Fos protein immunohistochemically, we used rats – intact ones and those exposed to a 48-hour osmotic stress, which received Epitalon intranasally, at a dose of 0.5 μg per animal (the last infusion was performed 2 h before extraction of the material). A photo-optic study of the pineal gland parenchyma was also performed in animals of all the groups.

c-Fos protein belongs to the activating triggers of the pineal synthetic processes in response to extreme factors, in particular to stress [Kovaks, 1998]. In our experiments, the nearly complete absence of c-Fos in the rat pineal glands was probably associated with the chronic type of the stress, since according to the published data, the maximum c-Fos content was usually observed 1–2 h after the stress impact. Chronic stress suppressed c-Fos gene activity in the paraventricular hypothalamic nucleus [Chen and Herbert, 1995] due to the high glucocorticoid blood level s [Kovacs and Sawchenko, 1997]. Besides, the nuclear glucocorticoid receptor and AP-1 transcription factor revealed a regulatory antagonism [Unlap and Jope, 1994].

A slight but statistically significant c-Fos increase was observed only in stress-exposed pinealocytes of rats after intranasal Epitalon infusions (fig. 12).

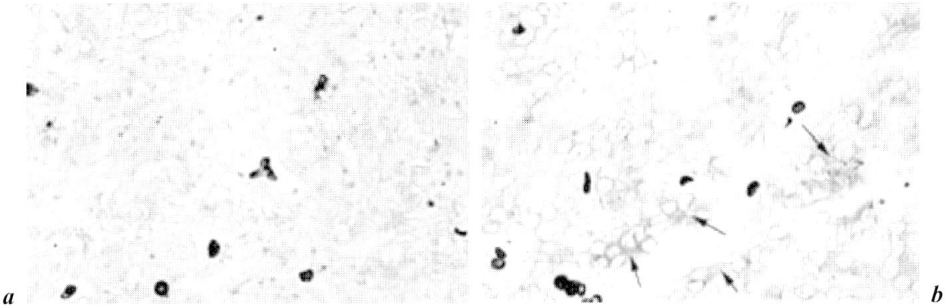

Fig. 12. Immunohistochemical detection of c-Fos protein. (***a***) Without Epitalon infusion. (***b***) With intranasal Epitalon infusion (cell clusters producing c-Fos protein).

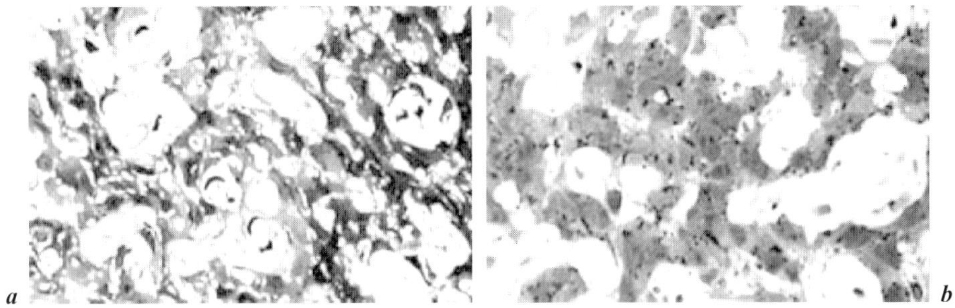

Fig. 13. Pineal parenchyma after a 48-hour water and food deprivation. (***a***) Without Epitalon infusion. (***b***) After intranasal Epitalon infusion.

The fact that Epitalon exerted this effect only in the case of stress, confirmed its participation in the specific mechanisms of pineal self-regulation under extreme conditions.

It was revealed that c-Fos in pinealocytes is not equally distributed. In stress, the clusters (5–10 cells) containing c-Fos in the cytoplasm were registered. Presumably, these clusters represent the very groups of interacting pinealocytes.

During pineal parenchyma morphometry, it was found that dilation of the capillaries and the pericapillary space induced by osmotic stress (fig. 13) was partially suppressed with intranasal Epitalon infusions.

Thus, Epitalon exerts an effect on the pineal gland in the case of an impact of extreme factors on the organism. Epitalon prevents osmotic-stress-induced pathologic changes in the pineal parenchyma structure. In addition, the effect of Epitalon on the pineal gland is probably mediated by activation of the protooncogenes.

5

Effect of Peptides on Carcinogenesis in Transgenic Mice

Genetically modified animals are used more and more often for studying the genetics of aging and participation of these or those genes in the development of age-related pathology, cancer included [Anisimov and Alimova, 2001; Anisimov, 2003]. Investigation of the effect of peptides on aging and carcinogenesis in transgenic animals is of high interest. Analysis of the mechanism of effect of peptides becomes very important due to the development of tumor biology therapy, aiming at prevention of metastasis and enhancement of protective functions of the organism [Gutterman, 1992; Paltsev and Ivanov, 1995].

The anticarcinogenic activity of Vilon (Lys-Glu) and Epitalon (Ala-Glu-Asp-Gly) was studied in transgenic female FVB mice bearing the mammary cancer gene HER-2/neu [Alimova et al., 2002; Anisimov et al., 2002; Khavinson, 2002]. The product of this gene is referred to the group of epidermal growth factor receptors. 80% of transgenic mice develop mammary adenocarcinoma. Also, early onset of age-associated disturbances of the estrous function and a decrease in life span are observed in such mice.

The transgenic mice, starting from 2 months age, were administered subcutaneously with either 0.1 ml of 0.9% NaCl solution (control) or Vilon or Epitalon (1 μg per mouse). The injections were made during 5 successive days monthly. The results are given in table 3. The number of mice with no evidence of mammary tumors was 1.6 times greater in the Epitalon-treated animals than in the controls. The number of single tumor-bearing mice amounted to 7% in the control group vs. 4% in the Vilon-treated animals and 16% in the Epitalon-treated group. The respective parameters for the mice with 2 or more tumors were 75, 85 and 56%, respectively. The maximum size of mammary adenocarcinomas in the Epitalon-treated mice was 33% less than in the control group.

Table 3. Effects of Vilon and Epitalon on the development of mammary adenocarcinomas and expression of the HER–2/neu oncogene transgenic mice

Index	Control	Vilon	Epitalon
Number of mice	28	27	25
Number of mice with mammary gland tumor	23 (82%)	24 (89%)	18 (72%)
Time of detection of the 1st tumor, days	168	150	188
Total number of tumors	106	114	85
Number of mice without tumors	5 (18%)	4 (15%)	7 (28%)*
Number of mice with one tumor	2 (7%)	1 (4%)	4 (16%)*
Number of mice with 2 and more tumors	21 (75%)	22 (81%)	14 (56%)**
Maximum tumor diameter, cm	2.1 ± 0.2	1.9 ± 0.2	1.4 ± 0.2*
Relative expression of mRNA of the HER–2/neu gene, conv. unit	1.64 ± 0.04	0.83 ± 0.10**	0.45 ± 0.21**

*p < 0.05, **p < 0.001 as compared with the control.

Expression of the mRNA of the HER-2/neu gene determined with real time PCR assay was 3.7 lower in tumors of Epitalon-treated mice than in tumors of the control mice. These data suggest that suppression of mammary carcinogenesis in transgenic mice by Epitalon is mediated by the inhibition of the HER–2/neu oncogene expression.

The above differences in the effects of different peptides indicate that these effects are unlikely to result from the organism's non-specific responses to peptide application. At the same time, studies at the level of the whole organism do not provide for unambiguous conclusions concerning the specific targets and mechanisms of their actions. Taking into account the parallelisms between the effects of Thymalin, Thymogen, and Vilon and those of Epithalamin and Epitalon, we can presuppose that the first three mainly affect the immune reactions, whereas a neuroendocrine component prevails in the effects of Epithalamin and Epitalon. Furthermore, possible interactions between the neuroendocrine and immune systems must also be taken into account. In particular, melatonin reveals an immunomodulatory activity. Stimulation of melatonin production by Epithalamin and Epitalon make it plausible that the in vivo effects of Epitalon in particular on HER-2/neu gene expression are mediated by melatonin. Moreover, that antitumor effect of melatonin is especially pronounced in respect to hormone-dependent tumors, such as mammary adenocarcinomas.

6

Effect of Peptides on Chromosomes

6.1. Effect of Vilon and Epitalon on Number of Chromosome Aberrations in Corneal Epithelium and Bone Marrow Cells of Hypokinesic Rats

In order to estimate the mutagenic effect of Vilon (Lys-Glu) and Epitalon (Ala-Glu-Asp-Gly) on hypokinesia rats, their effect on the number of chromosome aberrations in cornea epithelium and bone marrow cells of rats was studied. The peptides were administered intraperitoneally to the animals for 5 days (last injection conducted 1 h before the onset of hypokinesia) at a dose of 2.5 μg/kg body weight.

Hypokinesia was modeled by putting the animals into cramped cells made of organic glass for 24 h with highly restricted mobility.

Hypokinesia is known to induce imbalance of the antioxidant and prooxidant systems indices, which determines the intensity pathologic processes [Khavinson et al., 2003a]. Oxidant reactions trigger disorders of cellular homeostasis regulation, facilitating the development of the disease and the processes of early aging [Harman, 1996; Yu and Yang, 1996]. Active oxygen forms and products of lipid peroxide oxidation can also induce gene and chromosome mutations.

Analysis of the number of chromosome aberrations in rat corneal epithelium showed that hypokinesia led to a 7-fold increase in chromosome aberrations, while the studied peptides reduced the development of this process (table 4). Moreover, it is Vilon that reveals a more pronounced anti-mutagenic activity.

A significant increase of chromosome aberrations in bone marrow cells was registered under the effect of hypokinesia.

Application of Vilon and Epitalon contributed to the reliable reduction of chromosome aberrations (table 5).

Table 4. Effects of Vilon and Epitalon on the number of chromosome aberrations in rat cornea epithelium in hypokinesia

Animal group	Number of anaphases	Number of chromosome aberrations	Chromosomes aberrations from total anaphases number (%)
Intact rats	700	4	0.6 ± 0.2
Control (hypokinesia)	700	29	4.2 ± 0.7*
Vilon + hypokinesia	1,000	4	0.4 ± 0.1**
Epitalon + hypokinesia	1,000	12	1.2 ± 0.3**

*p < 0.05 as compared to the intact animals index; **p < 0.05 as compared to the control index.

Table 5. Effects of Vilon and Epitalon on the chromosomes aberrations number in rats bone marrow cells in case of hypokinesia

Animal group	Number of anaphases	Number of chromosomes aberrations	Chromosomes aberrations from total number of anaphases (%)
Intact rats	700	9	1.3 ± 0.3
Control (hypokinesia)	700	34	4.8 ± 0.8*
Vilon + hypokinesia	700	6	0.9 ± 0.3**
Epitalon + hypokinesia	700	7	1.0 ± 0.3**

*p < 0.05 as compared to the intact animals index; **p < 0.05 as compared to the control index.

Thus, the anti-mutagenic effect of Vilon and Epitalon on hypokinesia confirms the protective properties of these peptide preparations.

6.2. Effect of Epitalon on the Incidence of Chromosome Aberrations in Senescence-Accelerated Mice

An age-related increase in the incidence of somatic mutations can result from the accumulation of irreparable DNA damage and underlie the development of age-related pathology, including malignant tumors. The dynamics of accumulation of these mutations with age differs in different mouse strains, which can determine differences in their life span, rate of aging, and the incidence

Table 6. Epitalon effect on chromosomes aberrations frequency in bone marrow cells of 12-month-old female SHR, SAMP-1 and SAMR-1 mice

Mice line; experimental series	Chromosomes aberrations frequency, %	Δ, %
SHR		
Intact	8.2 ± 0.41	
Control (physiological salt solution)	8.4 ± 0.38	
Epitalon	6.9 ± 0.45*	−17.9*
SAMR-1		
Control (physiological salt solution)	10.3 ± 0.11**	
Epitalon	7.2 ± 0.08*	−30.1
SAMP-1		
Control (physiological salt solution)	18.7 ± 0.51#	
Epitalon	15.0 ± 0.28*	−19.8

*p < 0.001 as compared to the control index; **p < 0.001 as compared to the SHR control mice index; #p < 0.001 as compared to the SHR and SAMR-1 control mice index.

of spontaneous tumors between different strains [Tucker et al., 1999; Vijg, 2000]. For example, SAMP mutant mice are characterized by a shorter life span (≤14 months) and exhibit signs of rapid aging compared to control SAMR mice [Odagiri et al., 1998; Takeda, 1999; Tucker et al., 1999].

Basing on the previously tested geroprotective properties of Epitalon (Ala-Glu-Asp-Gly) [Khavinson, 2002], a study was performed to elucidate the effect of long-term Epitalon treatment on the incidence of chromosome aberrations (CA) in female SAMP-1, SAMR-1 mice, and long-living SHR mice used in experimental oncology [Rosenfeld et al., 2002].

Starting from the age of 2 months, the mice were injected subcutaneously with Epitalon monthly (1 μg/mouse, 5 times a week). The number of CA in bone marrow cells in mice of all groups was analyzed at the age of 12 months. Bridges and fragments in cells during the anaphase were considered as CA.

The incidence of CA in rapidly aging SAMP-1 mice was 1.8-fold higher than in the SAMR-1 mice without signs of accelerated aging, while in this latter strain this parameter surpassed the corresponding value in age-matched SHR mice by 23%. Long-term Epitalon treatment significantly reduced the incidence of CA in mice of all three studied strains in comparison with age-matched controls. The most pronounced effect was observed in SAMR-1 mice (table 6).

Interestingly, the incidence of CA in Epitalon-treated SAMR-1 mice decreased even below the corresponding level in control long-living SHR mice of the same age.

SAMP mice are known to develop normally until the age of 4 months. After that they show increasing level of oxygen active forms, accompanied by the signs of rapid aging, including accelerated accumulation of CA compared to SAMR mice [Odagiri et al., 1998; Takeda, 1999]. Long-term treatment with Epitalon prolonged the life span of *Drosophilia melanogaster* and CBA mice, suppressed free radical processes, and inhibited the development of spontaneous tumors in mice [Khavinson et al., 2003a]. There is every evidence that Epitalon antimutagenic effect plays an important role in the realization of its geroprotective and antitumor effects.

Hence, our experiments showed that 12-month-old senescence-accelerated SAMP-1 mice are characterized by having a higher incidence of CA in bone marrow cells compared to normally aging SAMR-1 mice and long-living SHR mice of the same age. Long-term treatment with Epitalon markedly decreased the incidence of CA in SAMP-1, SAMR-1 and SHR mice.

6.3. Effect of Vilon and Epitalon on Nucleolus Organizers in Human Thymic Cells

It is well known that thymus peptides play a very important role in maintaining the structural and functional homeostasis of cellular populations [Morozov et al., 2000]. Thymus peptides reveal thymomimetic activity and regulate both the thymic inner antigen independent stage of T lymphocyte differentiation and the immunocompetent peripheral T cell differentiation. At the same time, T cells play the leading role not only in the regulation of immunocompetent cell growth but also that of somatic cells.

Study of the peptide effect on ribosome formation, which is also regulated by argyrophil proteins of nucleolus organizers and on which the synthesis of the corresponding proteins takes place, is a matter of considerable interest. We studied the effect of the peptides Vilon (Lys-Glu) and Epitalon (Ala-Glu-Asp-Gly) on the expression of nucleolus organizer argyrophil proteins in thymocytes and epithelial cells which were jointly cultivated.

We also studied jointly cultivated human embryonic thymocytes and thymic epithelial cells obtained from abortion material at the 20th week of development. Vilon and Epitalon were used at concentrations of 2, 20 and 200 ng/ml. The peptides were introduced into the culture medium on the 2nd day. The study was conducted 24 h after introduction of the peptide. AgNOR-associated proteins were stained in the cells by silver nitrate. Incubation time

Table 7. Silver granules content in thymocytes nuclei and thymic epithelial cells in case of their combined cultivation under the effect of different Vilon concentrations

Experiment	Thymocyte nuclei	Thymic epithelial cells nuclei
Control	3.15 ± 0.08	22.07 ± 0.47
2 ng/ml	3.30 ± 0.13	24.90 ± 0.41*
20 ng/ml	3.73 ± 0.12***	27.80 ± 0.91**
200 ng/ml	3.93 ± 0.10***	28.80 ± 0.59**

*p < 0.05; **p < 0.01; ***p < 0.001 as compared to the control.

for argyrophil protein staining in thymocytes was 15 min, and in epithelial cells it was 10 min (taking into consideration the different expression of AgNOR-associated proteins in thymocytes and thymic epithelium cells). The intensity of argyrophil protein expression was evaluated according to the number of silver granules which appeared during the reaction. The granules were counted in 100 cells (s.g. 1,800). We estimated the average number of silver granules per 1 thymocyte and epithelium cell nucleus and thus the average number of granules in each group.

The introduction of Vilon into the culture medium at a concentration of 2 ng/ml did not provoke any changes in the cellular composition of thymocytes (table 7). The increase of Vilon concentration in the culture medium to 20 and 200 ng/ml led to an increase in the number of silver granules in thymocytes. But there was no significant difference in the number of silver granules when the Vilon concentration was 20 and 200 ng/ml (table 7). At the same time, with these Vilon concentrations (20 and 200 ng/ml) we registered a significant morphologic heterogeneity of the thymocyte population. An increase was observed in the number of large cells with delicate nucleus chromatin structure, which were morphologically similar to the blast-transformed lymphocytes, which appear during antigenic stimulation and under the effect of mitogenes. In the blast-transformed cells, we registered a higher content of silver granules than in the thymocytes.

Epithelial cells showed an increase in the number of silver granules under all concentrations of Vilon. Like in the thymocytes the most significant changes were observed with 20 and 200 ng/ml of Vilon (table 7).

Epitalon suppressed argyrophil protein expression in the thymocytes. At the same time, its effect did not depend on concentration (table 8). Epithelial cells revealed a decreased number of silver granules as compared with the control, which was similar under all concentrations of Epitalon (table 8).

Thus, Vilon stimulated the expression of the argyrophil nucleolar organizer region associated proteins in thymocytes and epithelial cells, which are

Table 8. Silver granules content in thymocytes nuclei and thymic epithelial cells in case of their combined cultivation under the effect of different Epitalon concentrations

Experiment	Thymocytes nuclei	Thymic epithelial cells nuclei
Control	3.15 ± 0.08	22.07 ± 0.47
2 ng/ml	2.42 ± 0.09***	20.50 ± 0.51*
20 ng/ml	2.52 ± 0.09***	20.70 ± 0.51**
200 ng/ml	2.47 ± 0.13***	20.90 ± 0.56***

*$p < 0.02$; **$p < 0.01$; ***$p < 0.001$ as compared to the control.

responsible for the synthesis, gathering and transportation of ribosomes into the cytoplasm, predetermining the intensity of protein synthesis which takes place in these structures. Epitalon reduced the argyrophil nucleolar organizer region associated proteins expression exerting an opposite effect on the processes of ribosomal generation and thus, on the intensity of protein synthesis in these cells. Vilon introduction into the cultural medium caused an intensive generation of large blast cells among the thymocytes. This transformation of lymphocytes into blast-proliferative cells, with the signs of intensive metabolic activity, is considered to take place in case of their stimulation by mitogenes. One of the earliest processes, observed under mitogenic stimulation, is exactly an increase of the functional activity of argyrophil nucleolar organizers and ribosomal genes localized in them that is accompanied by the increased speed of ribosomal RNA synthesis [Derenzini et al., 1995].

Presumably, high expression of the argyrophil nucleolar organizer region associated proteins in cells and the reaction of the blast transformation of thymocytes which are observed when Vilon is introduced into the cultural medium results from the peptide direct mitogenic effect and reflect a high transcriptional activity of the ribosomal genes that can evidence the inclusion of the peptide into the mechanisms of nucleolar organizer regions and ribosomal genes functional regulation. High argyrophil protein expression in thymocytes and epithelial cells after the introduction of Vilon into the culture medium can also evidence stimulation of the cellular proliferative activity, which correlates with literature data reflecting the direct dependence of cellular proliferative activity on the intensity of argyrophil protein expression [Sirri et al., 1995]. A positive correlation between the number of argyrophil proteins, cell growth fracture and cell proliferation speed is typical of the proliferating cell population.

Epitalon, unlike Vilon, provoked a decrease in the argyrophil protein expression level in thymocytes and to a lesser extent in epithelial cells; this can be considered as a sign of suppression of the ribosomal genes transcriptional activity as well as suppression of the cellular proliferative activity. Besides a decreased level of argyrophil protein expression during the introduction of Epitalon into the culture medium can evidence the emergence of highly differentiated slowly proliferating cells.

6.4. Effects of Small Peptides on Blood Lymphocyte Chromatin in Senile Humans

The therapeutic potential of peptide bioregulators is manifested by stimulation of immune and reparative processes, reduced probability of disease and accelerated aging development [Morozov et al., 2000; Khavinson, 2002; Korkushko et al., 2002; Khavinson and Anisimov, 2003; Khavinson et al., 2003a]. Peptide bioregulators induce metabolic changes controlled through gene regulation mechanisms realized in chromatin domains – 'functional units' of heredity. Functional structural domains regulate the interrelated processes: genetic activity, time of reproduction, and degree of chromosome condensation.

Structural condensation of chromosomes closely correlates with functional heterogeneity. Compactly condensed heterochromatin, heterochromatinized (condensed eu- and heterochromatin) chromosome regions are genetically inactivated and their replication is delayed. Decondensed (euchromatin) chromatin regions actively function. An obligatory condition for the transcription activity of genes is active chromatin [Lundgren et al., 2000]. Aging is associated with intense heterochromatinization paralleled by the inactivation of previously active genes [Lezhava, 2001]. Heterochromatinization of the chromosome regions is conditioned by the cycle lability and is a reversible process [Khavinson et al., 2002a].

Small peptides and peptidomimetics influence gene expression, which causes strong suppression of RNA synthesis, as well as tumor growth and a significant activation of the metabolic pathways retarded by aging [Marsili et al., 2000].

In connection with this, we studied the effects of small peptides on the chromatin domain system in cultured lymphocytes from senile subjects, particularly their effects on chromatin reactivation – deheterochromatinization of heterochromatin satellite regions (activity of synthetic processes of ribosome genes), common, structural (variability of C-blocks of chromosomes 1, 9 and 16) and facultative heterochromatin [Khavinson et al., 2004b].

The study was carried out on chromosomes of 150 PHA-stimulated and 40 PHA-nonstimulated cultures of lymphocytes from 95 healthy individuals aged 75–88 years (control: 25 donors aged 20–40 years). Two cell cultures from each individual were studied: intact and stimulated with Vilon (Lys-Glu), Epitalon

(Ala-Glu-Asp-Gly), Livagen (Lys-Glu-Asp-Ala), Prostamax (Lys-Glu-Asp-Pro), and Cortagen (Ala-Glu-Asp-Pro) for comparing test results with the control from the same individual in each case. The final concentration of Epitalon and Livagen was 0.005 μg/ml and that of Vilon, Prostamax, and Cortagen was 0.01 μg/ml. Thirty to fifty metaphase plates of intact and peptide-treated lymphocyte cultures (10 cultures in each group) were analyzed for each parameter. Studies by differential scanning microcalorimetry (DSM) were carried out on 40 (20 intact and 20 peptide treated) PHA-nonstimulated 22-hour cell cultures. The DSM method is based on the fact that thermostability differs in different chromatin fractions.

Activity of ribosomal genes in intact and peptide-treated cultures was evaluated using Ag-binding and quantitation of associations of nucleolar organizer regions (NOR) of acrocentric chromosomes. Changes in activity were evaluated by comparing two binomial totalities.

The polymorphism of structural C-heterochromatin of chromosomes 1, 9 and 16 was evaluated. Comparative analysis of C-stained chromosomes in intact and peptide-treated cultures was carried out using a standard classification, according to which the size of the C-segments of chromosomes 1, 9 and 16 was compared to the short arm of chromosome 16 and the results were classified into 5 categories (a, b, c, d, e). The variability of facultative heterochromatin was evaluated by the incidence of sister chromatid exchanges (SCE) in intact and peptide-treated lymphocytes.

In situ hybridization and differential Ag staining showed that human ribosome genes involved in protein synthesis are located in the satellite regions of acrocentric chromosomes (in NOR) [Heliot et al., 2000]. Satellite regions participating in the nucleolus formation are predominantly heterochromatic areas.

The capacity of acrocentric chromosomes to form associations is determined by the presence of two chromatid satellite regions. Chromosomes are considered associated if a pair of their satellite regions is connected. The associative capacity of the satellite regions positively correlates with the intensity of Ag staining, which, in turn, is determined by the activity of ribosome genes. One should remember that the absence of a satellite sequence (because of condensation) attests to the activation of ribosome genes [Olson et al., 2000]. The incidence of associations formed by silver-stained fragments of acrocentrics and the number of Ag-positive NOR per cell decreased considerably in senile subjects (because of heterochromatinization of the satellite regions) in comparison with medium-aged subjects [Lezhava, 2001].

Analysis of the incidence of Ag-positive NOR showed different argentophilia of certain acrocentrics in the control and peptide-treated cell cultures (table 9). The number of Ag-positive NOR of associative acrocentrics per cell in peptide-treated cultures was significantly higher in comparison with the control

Table 9. Effect of small peptides on lymphocyte chromatin in senile subjects

Variant of experiment	Associating acrocentric chromosomes (per cell)	Deheterochromatinisation of facultative heterochromatin – SCE (per cell)	General heterochromatin	Structural heterochromatin (C-blocks)		
				Chromosomes		
				1	9	16
Control (20–40 years)	1.33 ± 0.06	7.7 ± 0.4	stable condition	stable condition	stable condition	stable condition
Control (75–88 years)	1.17 ± 0.05*	5.9 ± 0.2*	heterochromatinisation	heterochromatinisation	stable condition	stable condition
Vilon	2.39 ± 0.11**	9.9 ± 0.6**	deheterochromatinisation	heterochromatinisation	stable condition	stable condition
Epitalon	2.32 ± 0.12**	8.4 ± 0.5**	deheterochromatinisation	deheterochromatinisation	deheterochromatinisation	stable condition
Livagen	2.49 ± 0.14**	9.2 ± 0.4**	deheterochromatinisation	deheterochromatinisation	deheterochromatinisation	stable condition
Prostamax	2.50 ± 0.15**	12.0 ± 0.8**	deheterochromatinisation	deheterochromatinisation	stable condition	stable condition
Cortagen	2.20 ± 0.7**	10.1 ± 0.7**	deheterochromatinisation	heterochromatinisation	stable condition	stable condition

*$p < 0.05$ as compared to the control index (20–40 years); **$p < 0.001$ as compared to the control index (75–88 years).

cultures derived from senile subjects. It is noteworthy that all 5 peptides promoted a steady increase in the incidence of DD-, DG- and GG-type associations.

These results are in line with the published data. It was reported, for example, that hormones and some growth factors promoted deheterochromatinization (decondensation) of chromosomes in senile subjects as well, which resulted in increased transcription activity of the nucleolus-forming regions [Khavinson et al., 2002a].

The increased number of Ag-positive NOR and the high incidence of associations of acrocentric chromosomes in lymphocyte cultures treated with peptides indicates deheterochromatinization of the satellite regions, causing activation of the ribosomal genes during aging. It is noteworthy that the expression of ribosomal genes is regulated by the ATRX gene (member of SNF2-helicase/ATPase family) through modulation of the structure of satellite regions in acrocentric chromosomes [McDowell et al., 1999].

As it was found during microcalorimetric studies of cells and cellular nuclei heat denaturing that membranes, cytoplasmic structures, and nuclear proteins are denaturated at 40–70°C, while chromatin is denaturated at higher temperatures (74–107°C). The curves reflecting heat absorption by cultured lymphocytes without PHA stimulation and in peptide-treated cultures were characterized by intricate profiles of independent peaks. Endotherms I, II and III corresponded to chromatin denaturing. Peptides added to lymphocyte culture caused changes in the profiles of heat absorption curves. For example, addition of Vilon (fig. 14) shifted endotherms II and III towards lower temperatures by 2.9° and 1.0°C. Moreover, heat redistribution between transition stages II and III was observed: heat increased at stage II and decreased at stage III. Proceeding from published data, we hypothesized that the transition stage (endotherm) II was associated with uncoiling of 10- and 30-nm fibrils, while transition stage III was associated with uncoiling of loops consisting of 30-nm fibrils fixed to the nuclear matrix. These results indicate that Vilon causes heat redistribution between stages II and III, which is due to partial despiralization of loops to the level of 30-nm fibrils.

A similar picture was observed in the analysis of heat absorption curves after treatment with Epitalon, Livagen, Prostamax and Cortagen, i.e. endotherms II and III were also shifted towards lower temperatures.

Hence, the peptides Vilon, Epitalon, Livagen, Prostamax, and Cortagen unfold higher levels of chromatin organization in the lymphocytes of senile subjects, deheterochromatizing total (facultative and structural) heterochromatin.

Data reflecting the heteromorphism of structural heterochromatin (C-segments) separately, concerning chromosomes 1, 9 and 16 in lymphocytes and those treated with the studied peptides, in senile subjects, are presented in table 9.

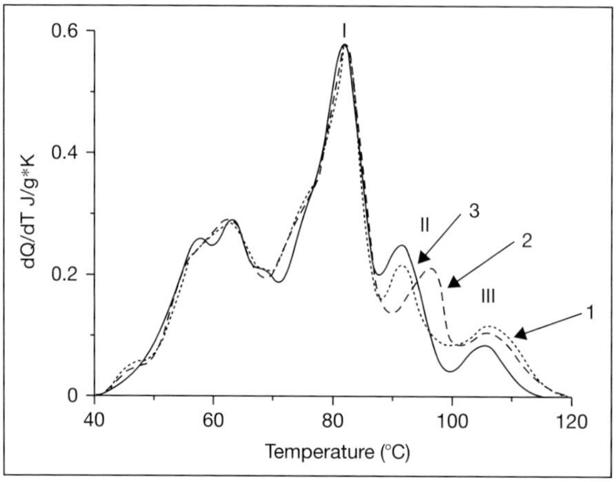

Fig. 14. Calorimetric curves reflecting lymphocyte chromatin denaturation process in elderly people. 1 = Intact lymphocytes in young people (26–35 years); 2 = intact lymphocytes in elderly people (76–81 years); 3 = Vilon-treated lymphocytes in elderly people (76–81 years). Endotherm regions: I = 80.3°C; II = 95°C; III = 106.2°C.

The results reflecting variability in the incidence of large (d and e) and small variants (a and b) of C-segments in certain chromosomes are heterogeneous for the tested peptides. In particular, chromosomes 1 and 9 were heteromorphic under the effects of Epitalon and Livagen, while Prostamax had such an effect only in chromosome 1. The degree of heteromorphism (decrease of the size of large blocks) was statistically significant for these chromosomes. After treatment with Vilon and Cortagen large and small variants of C-segments were recorded in chromosomes 1, 9 and 16 with the same frequency as in intact cells.

Chromosomes 1 and 9 are characterized by a pronounced variability of the absolute and relative sizes of C-heterochromatin in some diseases and during exposure to chemicals, while the distribution of heterochromatin C-segment variants in chromosome 16 remains stable [Khavinson et al., 2002a]. In addition, elongation of the C-segments of chromosome 1 (heterochromatinization of pericentromeric heterochromatin) was noted in senile subjects [Lezhava, 1999].

The results suggest that peptide treatment decreased the size of C-blocks in chromosomes 1 (Epitalon, Livagen, and Prostamax) and 9 (Epitalon, Livagen), which indicates the selective capacity of these drugs to induce decondensation (deheterochromatinization) of structural heterochromatin in the cells of senile subjects.

The results of the study of the effect of peptides on SCE incidence in lymphocyte cultures from senile subjects are presented in table 9. The incidence of SCE is usually considered an indicator of the reaction of cell homeostasis to endogenous and exogenous factors. This parameter can exhibit appreciable variety in different individuals, and therefore individual controls were used for testing each substance.

All the studied peptides increased the incidence of SCE in cells from senile subjects (table 9). The differences between control and experimental values were significant for all variants.

The effects of the tested bioregulators on the incidence of SCE varied for different groups of chromosomes. For example, Vilon and Livagen increased the number of exchanges for chromosome groups A, B, C, D, E and G; Prostamax was active towards groups A, B, C, D and G; Epitalon increased the number of exchanges for chromosome groups A, C, D and G, and Cortagen for groups A, C, and D. All these changes were significant. These data attest to a selective effect of each bioregulator on certain chromosomes.

No exchange processes are observed in the heterochromatin and heterochromatinized regions of chromosomes, i.e. SCE form mainly in euchromatin regions. Therefore, the detected increase of SCE values in lymphocytes of senile subjects is due to decondensation of the chromosomal heterochromatinized euchromatin regions under the effects of the studied peptides.

Hence, small peptides (Vilon, Epitalon, Livagen, Prostamax, and Cortagen) promote a 'release' of genes which are repressed because of heterochromatinization of the chromosomal euchromatin regions associated with aging.

7

Effect of Epitalon on Telomerase Activity, Telomere Elongation and Proliferative Potential in Human Somatic Cells

The results of performed studies point out the wide range of the Epitalon (Ala-Glu-Asp-Gly) geroprotective mechanisms which are manifested in decreased incidence of chromosome aberrations in rapidly aging SAM mice [Rosenfeld et al., 2002], the prolonged mean and maximum life span, and the reduced frequency of malignant tumor development in these animals [Anisimov et al., 2001, 2002a,b].

The telomerase theory of aging attributes the age-associated decrease of the tissue proliferative potential to critical shortening of telomeres in the course of cell division [Wright and Shay, 2002]. Telomerase is a ribonucleic enzyme constructing the telomere repeats TTAGGG, which are lost because of asymmetry in the synthesis of the leading and lagging DNA strains and is encoded by two genes: for RNA component and protein component of the enzyme, respectively [Hiyama et al., 1995]. In humans, the expression of telomerase protein component and the corresponding enzyme activity are observed in the majority of malignant, sex, early embryonic, and, presumably, stem cells. Human somatic cells have no telomerase activity [Meyerson et al., 1997].

In this experiment, we investigated the effect of Epitalon on the expression of the telomerase catalytic subunit, telomerase activity, and telomere elongation in human somatic cells [Khavinson et al., 2003b].

In this study, we used telomerase-positive HeLa culture and primary culture of human fetal lung fibroblasts 602/17 (24 weeks). Starting from passage 27, fetal fibroblasts were treated with Epitalon at a concentration of 0.05 μg/ml for 4 days and then analyzed.

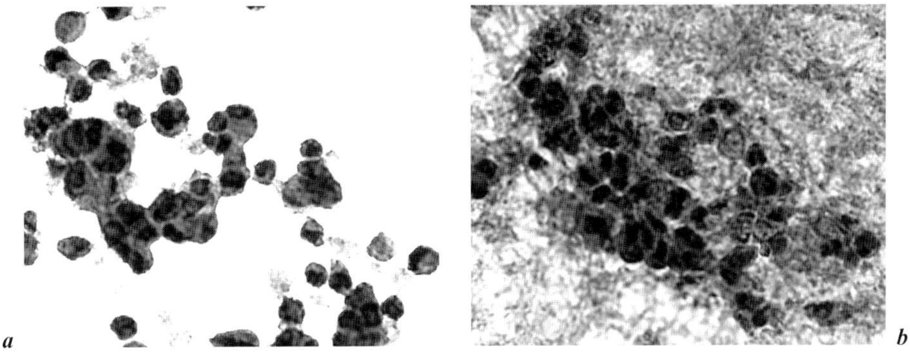

Fig. 15. HeLa cells (***a***) and Epitalon-treated fetal fibroblasts (***b***). Staining performed according to immune peroxidase method. ×400.

Immunohistochemical staining was carried out using murine monoclonal antibodies to the human telomerase catalytic subunit NCL-hTERT (Novocastra Laboratories). The universal peroxidase ABC kit (Dako) served as the second antibodies. Microscopy was carried out under a Nikon Eclipse E400 device with an AST1 digital camera.

Telomerase enzymatic activity was evaluated according to the protocol of telomere repeat amplification (TRAP) [Hiyama et al., 1995]. For in vitro analysis of telomerase activity, 2 μl of cell supernatant was added to 46 μl 1 Taq buffer containing 2 μ*M* TS primer (5′-AAT CCG TCG AGC AGA GTT-3′), 50 μ*M* dNTPs, and incubated at 30°C for 30 min. Then 2 μ*M* CX primer (5′-CCC TTA CCC TTA CCC TTA CCC TTA-3′), 5 U Taq-DNA polymerase (Promega) were added, and 35 PCR cycles with the following parameters were carried out: 1 min at 94°C, 1 min at 50°C, 1.5 min at 72°C on a DNA Engine PTC-200 thermocycler with warming cover (MJ Research). PCR products were separated electrophoretically in 10% nondenaturing polyacrylamide gel stained with SYBR Green (Molecular Probes) and visualized in UV light.

The mean telomere length in individual cells was measured by fluorescent in situ hybridization with flow cytofluorometry (flow-FISH) [Rufer et al., 1998].

The immunohistochemical study showed intense staining of telomerase-positive HeLa cells (fig. 15a) used as the positive control, and fetal fibroblasts treated with Epitalon (fig. 15b). No such staining was seen in intact (control) fetal fibroblasts.

No telomerase activity was detected in intact fetal fibroblasts either, while HeLa cells and Epitalon-treated fibroblasts revealed it (fig. 16).

Fluorescent in situ hybridization with flow cytofluorometry demonstrated increased mean and maximum lengths of telomeres in G1 fetal fibroblasts

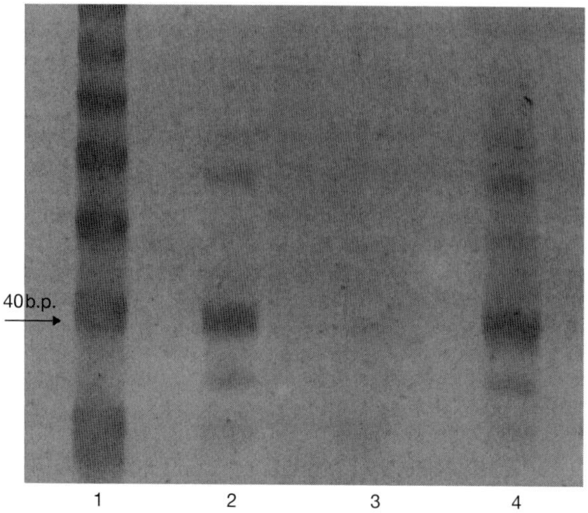

Fig. 16. Electrophoretic division and detection in PAAG of PCR-amplified elongated TS primers. 1 = 10 p.o. DNA marker; 2 = HeLa cells; 3 = fibroblasts; 4 = Epitalon-treated fibroblasts.

treated with Epitalon in comparison with control cells (fig. 17). The mean length of telomeres in G1 control cells was 180 CU. Epitalon increased the mean telomere length to 240 CU ($p < 0.05$).

These results indicate that in human somatic cells Epitalon can induce the expression of telomerase enzyme component, telomerase activity, and telomere elongation (by on average 33.3%). Hence, we demonstrated the induction of telomerase activity with a peptide for the first time. These data largely explain the geroprotective effects of Epitalon in various experimental models.

It is known that critically short telomeres cannot prevent chromosome fusion, which can lead to activation of protooncogenes and malignant transformation [Elmore and Holt, 2000]. Thus, activation of telomerase and telomere elongation under the effect of the peptide explains the mechanism of the anti-tumor effect of Epitalon in old animals.

Based on the received results, the next series of experiments were conducted for studying the effect of Epitalon on the proliferative potential of human fetal fibroblasts [Khavinson et al., 2004a].

The primary culture of human fetal lung fibroblasts 602/17 (24 weeks) was again used in the study. The cells were cultivated with 1:2 seeding once in 7 days up to passage 34, after which there was no further fibroblast division. For some fetal fibroblasts, beginning with passage 28, Epitalon was added into the culture medium at a concentration of 0.05 μg/ml. The mean telomere length in

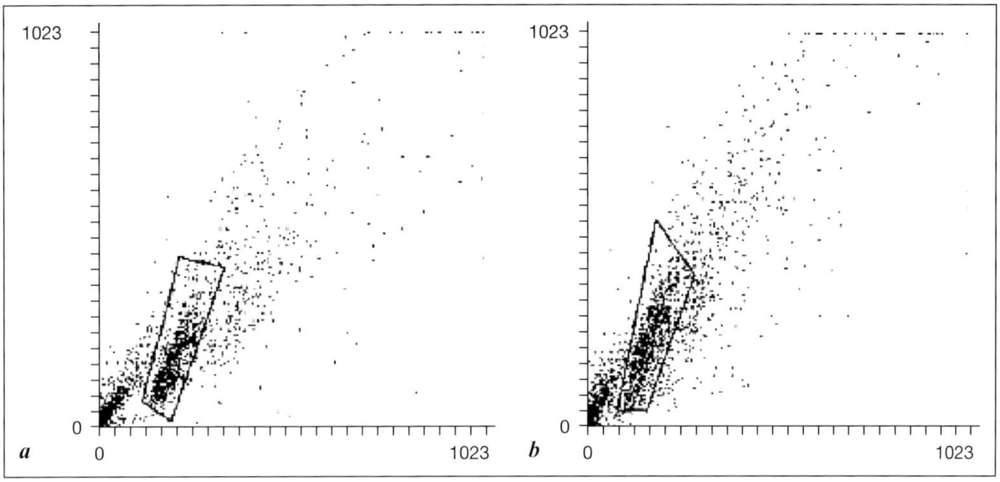

Fig. 17. Telomere length identification results using the in situ fluorescent hybridization with flow cytofluorometry method (flow-FISH). Fetal fibroblasts in control (***a***) and Epitalon-treated fetal fibroblasts (***b***). Propidium iodide staining of cell DNA (linear scale is given on the abscissa and telomere staining by PNA-FITC assay (linear scale) on the ordinate. G1 cellular cycle phase cells are framed.

individual cells was measured using the in situ fluorescent hybridization method with flow cytofluorimetry (flow-FISH).

During the study, 602/17 primary lung fibroblasts were found to demonstrate the loss of proliferative potential and cessation of cell growth on passage 34. In situ fluorescent hybridization with flow cytofluorimetry showed a decrease of mean and maximum telomere length in the G1 cellular cycle phase in 'old' fetal 602/17 fibroblasts after passage 32 (fig. 19a, b) if compared with the control 'young' cells subjected to 10 passages (fig. 18a, b). Mean telomere length in the G1 phase of the cellular cycle in the control (10 passages) made 20.9 ± 6.6 CU (fig. 18a, b). The long-term cell cultivation (32 passages) led to a decreased mean telomere length up to 10.9 ± 4.3 CU ($p < 0.05$) (fig. 19a, b). During propidium iodide staining of 'young' cells the mean DNA content in fibroblasts reached 13.2 ± 1.5 CU (fig. 18a, c), while in the 'old' cells this index reached 18.6 ± 1.9 CU ($p < 0.05$) (fig. 19a, c). This fact is, evidently, due to mass chromosome aberrations, which, in turn, were caused by the shortening of telomeres up to the critical level when they are no longer able to prevent chromosome fusion [Elmore and Holt, 2000]. According to flow-FISH results, the addition of Epitalon into the cultural medium led to an increase in mean telomere length in the G1 phase of 38 passage 'old' fibroblasts cellular cycle up to 26.2 ± 7.9 CU (fig. 20a, b), which significantly ($p < 0.05$) exceeds

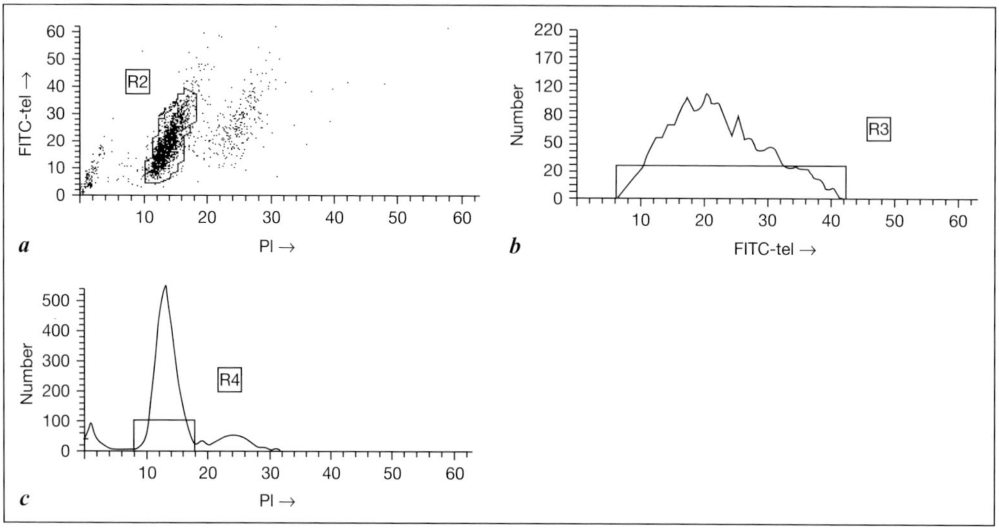

Fig. 18. 'Young' fetal fibroblasts of the10th passage. (***a***) Telomere length identification results using the in situ fluorescent hybridization with flow cytofluorometry method (flow-FISH). Propidium iodide staining of cell DNA (linear scale) (abscissa) and telomere staining by peptide-nucleotide assay (linear scale) (ordinate). G1 cellular cycle cells are outlined. (***b***) Histogram of cell distribution according to telomere length in the G1 phase of the cellular cycle. (***c***) Histogram of cell distribution according to DNA content in the G1 phase of the cellular cycle.

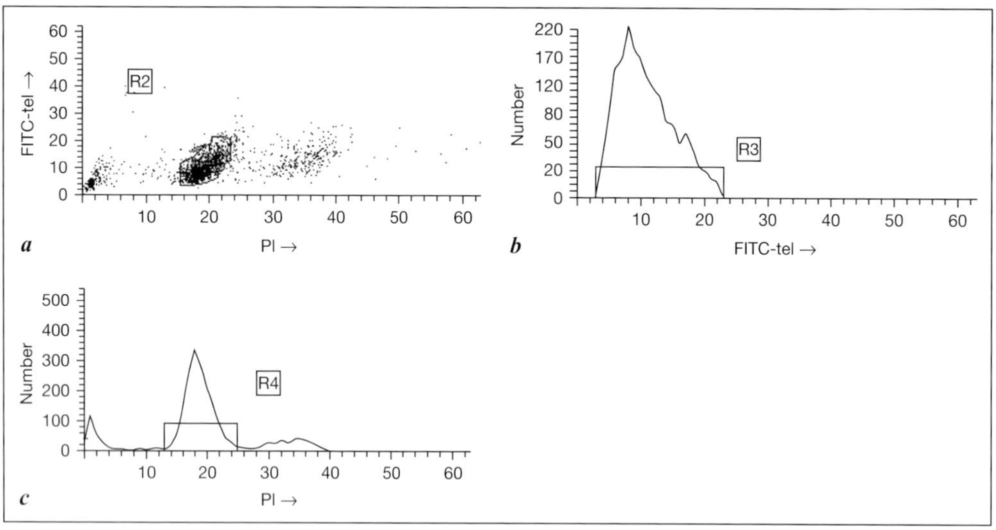

Fig. 19. 'Old' fetal fibroblasts of the 32nd passage. (***a–c***) As in legend to figure 18.

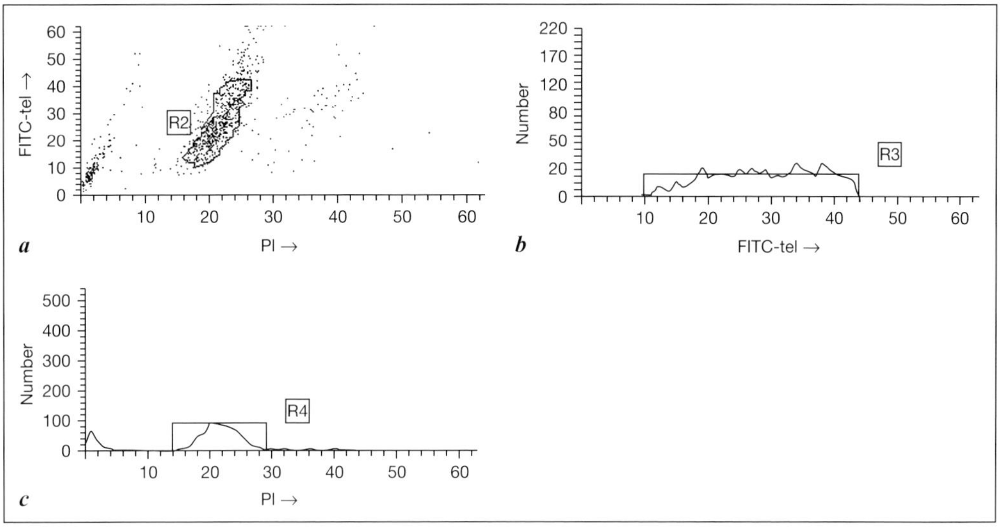

Fig. 20. Epitalon-treated fetal fibroblasts of the 38th passage. (***a–c***) As in legend to figure 18.

the mean telomere length in passage 32 cells (fig. 19a, b). In Epitalon-treated culture medium, the fibroblasts passed 44 divisions, and this process is ongoing. It is noteworthy that alongside with telomere elongation the DNA content in cells remained the same as in the cells that underwent long passing (fig. 19a, c, 20a, c). This, evidently, is connected with irreversibility of chromosome aberrations that already took place.

Thus, Epitalon contributes to prolongation of the life cycle of diploid human cells due to the overcoming of Hayflick's limit. These data give an explanation of the geroprotective effect of Epitalon on different experimental models.

8

Effect of Peptides on DNA Synthesis

8.1. Effect of Vilon on DNA Synthesis in Rat Thymus Cells under the Influence of Nitrites

Widespread modern technologies have led to an increased risk of environmental pollution, as well as to inevitable growth of the number of persons affected by adverse factors. Different factors of physical, chemical and biological nature, depending on the duration or intensity of their effect on the human organism, are able to exhaust adaptive and compensatory mechanisms and to cause deep-going disorders in different links of the immune protection system. This is caused by extremely complex and multicomponent processes of the formation and realization of defense reactions. The complex of structural and metabolic disturbances, related to the accumulation of DNA damage, modification of separate genes, alteration in DNA-protein bond durability, as well as chromosome modification, lie in the base of pathogenesis.

Constantly increasing chemical contamination of the environment is the prime reason for deterioration of the medicosocial situation, when nitrogen oxide compounds, in particular nitrates, are referred to as widespread ecologic toxins. Irrespective of whether nitrate compounds penetrate the organism or if they are reduced preliminarily, the toxic effect is manifested through the product of their reduction – the nitrites.

The character of the toxic effect of nitrites is manifested in two ways. Firstly, they are capable of binding with hemoglobin and producing methemoglobin, thus entailing certain hemoglobin characteristic changes – oxyhemoglobin dissociation process is disordered (2-fold suppression). Hemoglobin transport function is reduced after its transformation into methemoglobin or vice versa. Even after complete disappearance of methemoglobin, the oxygen capacity of the blood is reduced. Secondly, oxidoreduction reactions are being blocked at certain stages, and these toxicants cause a hypoxic state with corresponding manifestations in every organ. The increase in load of nitrogen oxide compounds on the human organism is an obvious reason not only for functional changes in internal organs and systems, but also for organ damage.

Moreover, nitrites tend to bind with secondary and tertiary amines, forming N-nitrosamines. The role of nitrates as predecessors of carcinogenic compounds has been confirmed experimentally.

On this basis, the study was performed regarding the effect of Vilon on the intensity of DNA synthesis in thymus cells of animals under the influence of sodium nitrite. In the course of the experiments, sodium nitrite was administered at a dose making 1/10 of LD_{50} (36 mg/kg). A crystalline substance was dissolved in distilled water ex tempore and introduced into the stomach by means of a nontraumatic metallic probe at the volume of 0.2 ml solution per 100 g animal's body weight. The toxicant was administered daily in the morning.

The developed intoxication model was used for evaluating the prophylactic and therapeutic efficiency of Vilon (Lys-Glu). For prevention, Vilon was administered at a dose of 1 μg intramuscularly, daily for 10 days, and then – in combination with sodium nitrite – for 10 or 20 days. For therapeutic purposes, Vilon was administered at the same dose for 10 or 20 days against the background of the effect of sodium nitrite.

On the 10th or 20th day of the experiment, animals from the corresponding groups were decapitated, and the content of high-molecular-weight and low-molecular-weight DNA was evaluated in tissue lysates of the thymus after the separation of intact chromatin and DNA fragments using the centrifuge method [Messmer et al., 1996].

The results of this study, displayed in table 10, show that in the case of intoxication with sodium nitrite, a fluctuation of total DNA content in the thymus was observed on the 10th day of the experiment, and on the 20th day its level decreased significantly. Meanwhile the share of low-molecular-weight DNA was significantly higher than in the control group – 3.4 times in 10 days and 4 times at the end of the experiment. The data obtained suggest that the toxicant's effect enhances the DNA fragmentation process in the thymus.

The introduction of Vilon into therapeutic and preventive schemes in the case of intoxication of rats with sodium nitrite, as a rule, exerted no significant positive effect on the DNA fractions' ratio in the thymus.

Also, the estimation of DNA content in the thymus of Vilon-treated rats revealed a new aspect of the thymotropic effect of the studied peptide. After 10 days of Vilon administration, a decrease was registered in the content of low-molecular-weight DNA, which was accompanied by an increase in organ weight. The received data prove enhancement of the proliferative processes and the suppression of cell death processes (possibly, apoptosis) in

Table 10. DNA content in rat thymus under the conditions of daily administration of sodium nitrite at a dose of 1/10 LD_{50} with and without Vilon at a dose of 1 μg

Animal group	DNA content in the thymus			
	summary DNA mg/g	high-molecular-weight DNA mg/g	low-molecular-weight DNA mg/g	low-molecular-weight DNA % of summary
Control (healthy animal)	12.69 ± 0.22	12.59 ± 0.49	0.099 ± 0.006	0.80 ± 0.06
Sodium nitrite 10 days	12.37 ± 1.35	12.20 ± 0.19	0.168 ± 0.006*	2.68 ± 0.17*
Sodium nitrite with Vilon 10 days	11.70 ± 1.18	11.52 ± 0.20 #	0.182 ± 0.048	1.59 ± 0.47
Vilon 10 days + sodium nitrite with Vilon 10 days	11.09 ± 0.19*	10.86 ± 0.19*#	0.222 ± 0.011*#	2.01 ± 0.11*#
Sodium nitrite 20 days	10.84 ± 0.52*	10.50 ± 0.56*	0.341 ± 0.012*	3.21 ± 0.19*
Sodium nitrite with Vilon 20 days	11.48 ± 0.16*	11.25 ± 0.22*	0.228 ± 0.013*#	1.99 ± 0.15*#
Vilon 10 days + sodium nitrite 20 days	10.80 ± 0.20*	10.57 ± 0.20*	0.236 ± 0.011*#	2.19 ± 0.14*#

*$p < 0.05$ as compared to the control index; #$p < 0.05$ as compared to the sodium nitrite group index.

thymus tissues under the effect of Vilon. By the end of the 20-day course of treatment, thymus weight still exceeded the control indices, but at this stage of the study an increased DNA content was revealed in the low-molecular-weight fraction, which, in our opinion, points out enhancement of the cell death process. This circumstance, on the one hand, requires further, more-detailed study of the mechanisms of the observed phenomena, and on the other hand, from a practical viewpoint reveals the necessity of strict regulation of the terms of treatment.

8.2. Effect of Vilon on DNA Synthesis in Blood Leukocytes of Old Rats Exposed to Small-Dose Radiochemical Impact

It is known that young pubertal rats (3–4 months old) exposed to small doses of irradiation and mercury show early hematological changes (mainly, lymphopenia). These changes are accompanied by genetic instability without corresponding activation of the matrix activity of leukocyte DNA. By the end of a restoration period – 30 days after such damaging influences – normalization of the quantity of lymphocytes in the blood was observed, as well as of the content and structure of leukocytic DNA. Taking into account earlier conducted studies [Ivanov et al., 1999], these results enabled to make an assumption stating that a decreased life span may be observed in the group of animals subjected to radiochemical influence in small doses and manifesting no increase in the matrix activity of DNA. The subsequent observation of actual survival of the animals after repeated radiochemical exposures at the same doses confirmed this theory.

The administration of Vilon at a dose of 5 μg/kg to young rats exposed to small doses of radiation and mercury enhanced leukopenia at the early stages due to granulocytes, and, to a lesser extent, due to lymphocytes. In this case, genome instability was also revealed, but it was accompanied by changes in the DNA structure, due to the increase in its matrix activity. As a result, 30 days later the hemocytological parameters (except the quantity of lymphocytes) and the indices reflecting the content and structure of DNA in blood leukocytes were normalized. This enabled an assumption that Vilon at such a dose, increasing the matrix activity of DNA, is able to exert a certain geroprotective effect in old animals in the case of combined low-level radiochemical damage.

Control animals received $Hg_2(NO_3)_2$ mercury salt with water ad libitum, at a concentration of 1 μg/l with respect to the weight of metal for 1.3 months of the experiment and were subjected to irradiation at a dose of 25 cGr using the 'IGUR-1' gamma installation (^{137}Cs, dose power 41 cGr/min) 3 weeks after the beginning of the administration of mercury. Vilon was administered at the dose of 5 μg/kg, daily, intramuscularly, for 5 days beginning from the day of irradiation and for 5 days more after a 20-day interval.

In order to evaluate the effect of small doses of irradiation and mercury and that of Vilon on the changes in the content and structure of DNA in blood leukocytes, we used modified methods [Ivanov et al., 1999]. The leukocyte DNA index was calculated, as well as the coefficient of relative fluorescence.

During the early periods after irradiation – 24 h after γ-irradiation and the first administration of Vilon in the case of chronic mercury exposure, a decreased quantity of granulocytes was observed, as well as a tendency towards the decrease in the quantity of lymphocytes, which are known to be the most radiosensitive cells in the blood. At the same time a tendency was observed towards a decrease in the DNA index in the leukocyte blood population (approximately by 16%), as compared to this index in intact animals. Along with that, significant changes in DNA structure were found as determined by the increase (by 30%) of a relative fluorescence coefficient. As stated, the long-term consequences of early disturbances in genome stability induced by low-dose radioactivity and mercury influences in rats without corresponding activation of transcription and/or reparation (to judge from the absence of an increase in the relative fluorescence coefficient of leukocyte DNA) were represented by the decreased life span of the animals [Ivanov et al., 1999].

Basing on the received data, we can conclude that there are age-related peculiarities in the reactions of both old and young rats in the case of Vilon administration after the impact of low-dose radioactivity and mercury. The administration of Vilon to old rats after low-dose radioactivity and mercury has an influence on the early stages of disease and prevented granulocytopenia and a tendency towards lymphopenia. In this case, genome instability was also revealed, but along with that changes took place in the structure of DNA, called forth by the increase in its matrix activity. As a result, 30 days later hemocytologic parameters were practically normalized, as well as the indices reflecting the content and structure of blood leukocyte DNA. This enables us to suppose that Vilon exerts a geroprotective effect in old rats in cases of low-level combined radioactive and chemical damage.

8.3. Effect of Vilon on Mitotic Activity of Hepatocytes in Rats after Partial Hepatectomy

Partial hepatectomy with 2/3 liver removal is a common model for the study of liver regeneration in rats. The animals were subdivided into 3 groups: group 1 – healthy animals, group 2 – control (rats subjected to partial hepatectomy), group 3 – operated animals, which subcutaneously received Vilon (Lys-Glu) at 2 and 24 h after surgery, at doses of 0.1 μg per rat. The animals in groups 1 and 2 received NaCl solution in the same volume and following the same schedule.

Hepatectomized rats were killed with ether 32 and 96 h after surgery. Control group rats were killed at the same time. The rat liver specimens were fixed in formalin. After staining the preparations with hematoxylin-eosin the mitotic index was identified in liver cells, as well as the amount of polyploid cells which were in the S-phase of the cellular cycle (the quantity of dividing cells).

The study of mitotic activity of regenerating liver cells 32 h after partial hepatectomy showed that the quantity of mitoses and cells in the S-phase of the cell cycle becomes twice as high as in liver from healthy animal. These differences were insignificant in the case of administration of NaCl solution, while after Vilon injections a significant increase was observed in the quantity of mitoses, DNA-synthesizing cells and total quantity of dividing cells (table 11).

During study of the liver preparations 96 h after hepatectomy, it was found that both rats which had received NaCl solution and Vilon-treated rats showed a significant enhancement of hepatocyte mitotic activity. Comparison of the data from the experimental (third) and control (second) groups revealed that Vilon-treated rats showed a 2-fold quantity of mitoses as compared to the rats treated with physiological solution. The amount of cells in the S-phase of the mitotic

Table 11. Effect of Vilon on the number of dividing cells in rat regenerating liver tissue 32 and 96 h after partial hepatectomy (% of total liver cells)

Animal group	Experimental stages		Mitotic index	Cells, in the phase of DNA synthesis %	Total number of dividing cells
Healthy animals + physiological salt solution	–		0.682 ± 0.013	1.752 ± 0.463	3.403 ± 0.498
Control (partial hepatectomy) + physiological salt solution	32 h	before	0.431 ± 0.019	1.043 ± 0.127	1.474 ± 0.143
		after	1.364 ± 0.595	2.063 ± 0.474	3.427 ± 1.066
	96 h	before	0.417 ± 0.053	0.924 ± 0.091	1.342 ± 0.060
		after	2.012 ± 0.146*	3.417 ± 0.295*	5.429 ± 0.388*
Partial hepatectomy + Vilon	32 h	before	0.450 ± 0.067	0.870 ± 0.100	1.320 ± 0.159
		after	2.314 ± 0.461**	3.882 ± 0.839**	6.196 ± 1.279**
	96 h	before	0.294 ± 0.084	0.982 ± 0.141	1.276 ± 0.128
		after	4.846 ± 0.334*+	4.664 ± 1.315**	9.510 ± 1.609*#

*p < 0.001 as compared to the pre-operational index; **p < 0.05 as compared to the control animals index; +p < 0.001 as compared to the control animals group index; #p < 0.05 as compared to the control animals index.

cycle in experimental rats did not differ significantly from that of hepatocytes in the S-phase in the control group, although on the whole the amount of dividing cells 96 h after hepatectomy in the regenerating Vilon-treated rat liver was 75% higher than in the rats treated with physiological solution (table 11).

Hence, Vilon administration to rats enhanced the mitotic activity of hepatocytes 96 h after partial hepatectomy, which proves the acceleration of repair processes in the liver.

9

Interaction between Small Peptides and DNA Double Helix

Results of biological studies confirm the direct effect of small synthetic peptides on chromatin activation in differentiated tissues. At present, due to the development of pharmacogenomics, study of the mechanisms of interaction between substances of different nature and DNA is one of the priority directions in the development of effective medicinal substances [Evans and McLeod, 2003; Wang et al., 2004]. The construction of peptides with certain physical and chemical characteristics and biological properties is an advanced method of creation of physiologically active substances and potential medicinal preparations [Nandel et al., 1999; Ryu and Nam, 2000; Verrecchio et al., 2000]. The search for peptides with minimal length preserving their ability to recognize specific nucleotide sequences has been performed for a very long time [Grokhovsky et al., 1994a,b]. However, until now an obvious gap exists between numerous facts of the specific effects of small peptides on the activation of protein synthesis which is manifested at the level of the organism as a whole, and a limited character of concepts regarding the molecular mechanisms of selective binding of peptides with the regulatory segments of the DNA double helix. That is why it seems useful to analyze the physical and chemical bases of intermolecular interactions in these main components of chromatin. The recognition mechanism of a specific sequence of bases in the DNA structure by the peptide is stipulated by the composition of elementary links of these biopolymers: amino acids and nucleotides.

It is known that peptide chains include 20 amino acids differing in terms in their lateral group composition. Table 12 indicates their structures and main characteristics. As these data show, the lateral groups of amino acids (R) reveal a wide range of hydrophilic and hydrophobic properties, which determine their role in structuring the solvent according to hydration or to the clathrate mechanism. Moreover, the lateral groups of lysine and arginine show properties of

Table 12. Amino acids, coded DNA [according to: Leo et al, 1971; Zamyatnin, 1973]

No.	Name	One-letter sign	Mol. mass Da	Lateral group R	Hydrophobity of the lateral group (−G) kJ/mol	Volume of amino acid residue in water cm³/mol
1	Glycine	G	75	$-H$	0	36.1
2	Alanine	A	89	$-CH_3$	0.5	53.2
3	Valine	V	117	$-CH(CH_3)_2$	5.3	83.9
4	Leucine	L	131	$-CH_2-CH(CH_3)_2$	7.5	100.1
5	Isoleucine	I	131	$-CH(CH_3)-CH-CH_3$	7.5	100.1
6	Hystidine	H	155	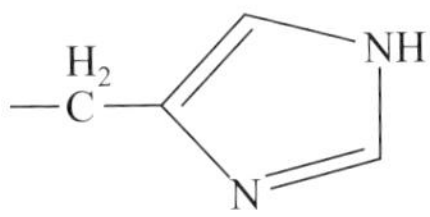	1.0	91.9
7	Proline	P	115	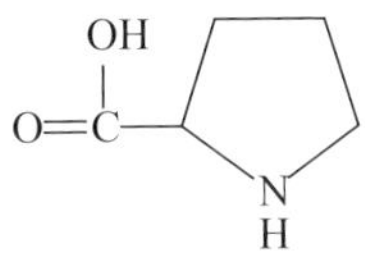	3.0	73.6
8	Thyrosine	Y	181	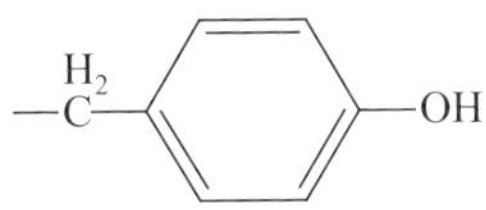	4.4	116.2
9	Phenylalanine	F	165	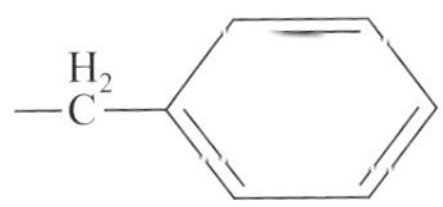	9.1	113.9
10	Tryptophane	W	204	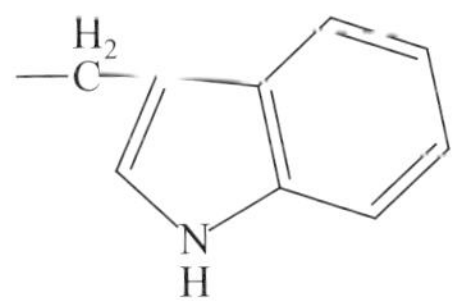	11.3	136.7
11	Serine	S	105	$-CH_2-OH$	1.3	53.4
12	Threonine	T	119	$-CH(OH)-CH_3$	5.8	69.7

Table 12 (continued)

No.	Name	One-letter sign	Mol. mass Da	Lateral group R	Hydrophobity of the lateral group (−G) kJ/mol	Volume of amino acid residue in water cm³/mol
13	Cystine	C	121	$-\overset{H_2}{C}-SH$	–	65.1
14	Metionine	M	135	$-\overset{H_2}{C}-\overset{H_2}{C}-S-CH_3$	5.9	97.7
15	Asparagine	N	132	$-\overset{H_2}{C}-\underset{\underset{O}{\Vert}}{C}-NH_2$	0.8	70.6
16	Glutamine	Q	146	$-\overset{H_2}{C}-\overset{H_2}{C}-\underset{\underset{O}{\Vert}}{C}-NH_2$	1.9	86.3
17	Asparagine acid	D*	133	$-\overset{H_2}{C}-\underset{\underset{O}{\Vert}}{C}-OH$	4.1	66.7
18	Glutamine acid	E*	147	$-\overset{H_2}{C}-\overset{H_2}{C}-\underset{\underset{O}{\Vert}}{C}-OH$	3.9	83.0
19	Lysine	K*	146	$-\overset{H_2}{C}-\overset{H_2}{C}-\overset{H_2}{C}-\overset{H_2}{C}-NH_2$	1.2	101.1
20	Arginine	R*	174	$-\overset{H_2}{C}-\overset{H_2}{C}-\overset{H_2}{C}-\overset{H}{N}-\underset{\underset{NH}{\Vert}}{C}-NH_2$	2.5	104.1

*In the proton-carrying form.

strong bases (pK_a values of 10.5 and 12.5, respectively), and the lateral groups of asparagine and glutamine acids possess pronounced acid properties (pK_a 3.6 and 4.2, respectively). Consequently, these lateral groups can take part in ion-ion and ion-dipole intermolecular connections.

The combination of proton-donor and proton-acceptor groups in the structure of amino acids provides for their participation in hydrogen bonds with other molecules, in particular, with solvent molecules.

Hydrophobic values of the lateral groups are indicated in table 12, i.e. the values of free energy of their relocation from a polar solvent (water) into a nonpolar one (octanol) show that all amino acids are to some extent hydrophobic, although for the majority of them the energy value of relocation into the nonpolar environment is not high and is commeasurable with heat energy – RT = 2.58 kJ/mol, the temperature being 37°C.

Measurement of the hydrophobic value of organic compounds has a long tradition, because already at the beginning of the 20th century pharmacologists found a correlation between hydrophobic value and physiologic activity of toxins, neuroleptics and narcotics. It was supposed that lipid structures of cellular membranes and myelin structures of nervous fibers constitute a nonpolar phase, where hydrophobic molecules are distributed with a high extent of selectivity. Later, Pauling [1961] studied the role of water hydration in the process of physiological intermolecular interactions. He showed that water is able to form complex compounds (clathrates) with hydrophobic groups of biologically active molecules, in particular, with analgesics. These compounds have a friable package and large molar volume. He also offered a hypothesis that combination of a polar (or electrically charged) group with hydrophobic parts of a molecule forms the basis of an interaction between analgesics and neural toxins on the one hand and cellular membranes on the other.

As indicated in table 12, amino acid residuals significantly differ in terms of partial molar volumes in the hydrated state. The highest values of molar volumes belong to the most hydrophobic amino acids: tyrosine, phenylalanine, and tryptophan, excelling the values of molar volumes of hydrophilic glutamine and asparagine acid residuals, lysine and arginine, carrying the electric charges on the lateral groups [Zamyatnin, 1973].

In the case of amino acids uniting into a peptide molecule, an individual hydrophilic to hydrophobic values ratio of the peptide's lateral groups is formed for every particular amino acid sequence [Alberts et al., 1994].

Spatial distribution of different electric charges and a dipole moment of peptide connection, which makes 3.5 D, determine the constant dipole moment and high polarization ability of the peptide. Due to the individual combination of amino acid residuals, these peptides present a wider array of energetic states than individual amino acids. But this is compensated by a decrease in spatial degrees of freedom in the system. Each of the simple molecules has 6 degrees of freedom for movement in the solution: three rotational and three progressive ones; there are 12 degrees of freedom for two non-interacting amino acids. In the case where they are united into one dipeptide, the degrees of freedom is reduced to 6. In terms of statistic thermodynamics, this is equivalent to an increased order in the system and to the corresponding decrease in entropy.

Figure 21 shows structural formulas of dipeptides revealing their geroprotective properties.

Carnosine and Thymogen combine hydrophilic and hydrophobic parts in their structure; in water solution these parts are located on different sides of the peptide bond plane. The hydrophilic groups of these dipeptides (left parts of the

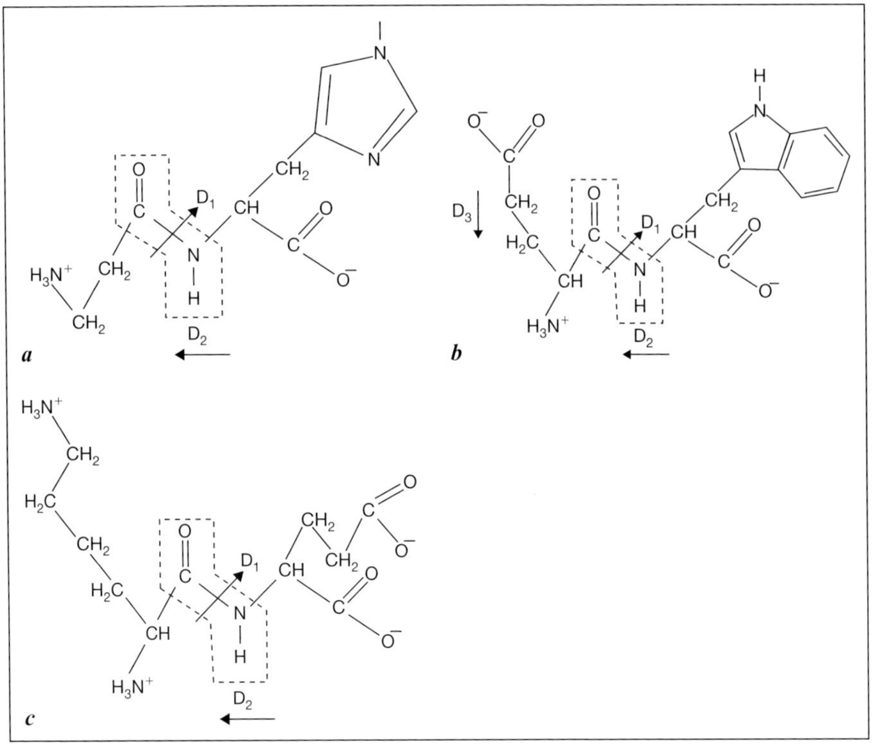

Fig. 21. Structural formulas of Carnosine (***a***), Thymogen (***b***) and Vilon (***c***). Boxed area – the plane of peptide bond, which is located on this page's plane. Arrows indicate dipole moments of D peptide bond and side groups. Local dipole moment vector is directed from negative charge to the positive one.

molecules shown in figure 21) differ in terms of charge: carnosine has a positively charged group, while Thymogen possesses a negatively charged carboxyl group. In the case where the residuum of glutamine acid in Thymogen is not in its pyro-form, its molecule has two local oppositely oriented dipole moments: between the C- and N-ends of the dipeptide (D_2) and between the α-amino- and carboxyl groups of the glutamine acid residuum. Thus, the summary constant dipole moment in Thymogen appears to be less than in the carnosine molecule.

A distinctive feature of the Thymogen molecule is that it combines the most hydrophilic (glutamil) and the most hydrophobic (tryptophan) amino acid residuals. The presence of tryptophan makes Thymogen different from all known peptide regulators of the thymic system and makes it close to the neural peptides of the neuromedin and liberin peptides classes. At the same time, Thymogen possesses all the immune modulating properties of Thymalin, but significantly surpasses the latter in terms of specific activity [Khavinson and Zhukov, 1992].

Presumably, the combination of a large hydrophobic lateral group of tryptophan and a compensated bipolarity facilitates the interaction of thymogen with hydrophobic segments of cellular membranes. In particular, it was shown that after the application of the EW dipeptide to the N-end of α-spiral polypeptide $(LSSLLSL)_3$, the elongated polypeptide binds with the surface of the cellular membrane, penetrates into its bilayer phospholipid structure, and generates ion canals in the membrane thus changing ion transport selectivity [Lear et al., 1997]. This is one of several examples of the fact that tryptophan residuals near the N-end of polypeptide chain increases the selectivity of the interaction between polypeptide and the external surface of the cellular membrane.

The dipeptide Vilon (Lys-Glu) is a structural link in many thymic hormones. As can be seen in figure 21, a distinctive feature of this dipeptide is the clear division of electrostatic charges between the two ends of the peptide: positively charged amino groups belong to the lysine residuals, while negative charges are concentrated at the glutamine acid residuals. Such structures confirm the ability of Vilon to take an active part in electrostatic (ion-ion and ion-dipole) interactions.

Summary of the hydrophobic values of the lateral groups of carnosine, Vilon and Thymogen, calculated according to the data presented in table 12 (1.5, 5.1 and 15.2 kJ/mol, respectively) show that the least hydrophobic, i.e. the most hydrophilic, is the carnosine molecule. Presumably, its hydrophilic value plays a certain role in the destabilization of hydrophobic amyloid β-structures.

The intermittence of amino acid residuals with positively and negatively charged lateral groups is characteristic for the majority of hydrophilic dipeptides revealing biological activity. For example, Lys-Asp and Glu-Arg are regarded as elementary regulators and stimulators of physiological function [Belokrylov et al., 1998]. Meanwhile, the range of functions which are influenced by these peptides is very wide and does not reveal tissue specificity. One can assume that these molecules not only reveal regulatory activity by themselves, but can also be agonists for other endogen regulators (including non-peptide ones), increasing cell receptor sensitivity to their action.

The above-described biologically active dipeptides have a comparatively simple structure and at the same time are able to take part in intermolecular interactions of different types. The elongation of peptide chain widens the range of molecule's functions and increases the scope of its regulatory functions. The examples are represented by synthetic tetrapeptides, the biological activity of which was presented in previous chapters in detail.

One ought to pay special attention to a principally new quality acquired by amino acids being united into a polypeptide chain. This quality is a complementary character of mobile conformations.

A common notion of complementarity refers to the categories of early concepts of natural philosophy. In the era of alchemy two descriptive concepts were formed: 'similar goes to similar' and 'oppositions tend to be close'. The first concept was based on the experience of division and purification of substances, and the second on the experience from chemical transformations, when the elements with opposite properties interact and complement each other. On the modern level of concepts regarding the composition of substances, the principle 'similar goes to similar' has been transformed into the theory of hydrophilic-hydrophobic interactions, and complementarity is illustrated as the attraction between positive and negative charges (in electrostatics), as the matching of convexo-concave surfaces (in geometry), and as the mutual correspondence of elements of the object and its mirror reflection (matrix complementarity in early polygraphy). In all these examples, the 'oppositions' are regarded as unchangeable objects.

Biological macromolecules, in particular peptides, preserve the complementarity of intermolecular interactions in a wide scope of conformations. This ability of theirs forms the base of catalytic activity of enzymes, and the model of the 'key-and-lock' complementary interaction was used for the first time in the early days of enzymology. Later, the same model was used during discussion of the specificity of antigen binding with antibody and of the selectivity of interaction between receptor and ligand.

In the course of study of the nucleic acids, the matrix complementarity of nucleotide pairs model was used for the first time. With its help, it was proven that these macromolecules control the precise copying of their own structure, because the initial macromolecule serves as the matrix for the production of the next one.

The matrix complementarity principle is very simple and elegant. Its point is that every element of orderly molecular construction (electrostatic charge, proton-donor or hydrophobic group) is in the same order copied on the other macromolecule, the elements of which are complementary to the elements of the first molecule.

A closer look revealed a matrix complementarity both in peptide synthesis and in the formation of phospholipid bilayer membranes, as well as in the self-assembly of polymers [Bredas, 1998].

Analysis of amino acid sequences is aimed at the rationalization of a vast amount of already available data on the primary structure of polypeptides. The basic method of this analysis is revealing the elements of similarity in structures of different proteins. The comparison of conservative chain segments with functional characteristics of the molecule enables many proteins in one group to be united. For example, the receptors related to G-proteins (muscarine, opioid, adrenal receptors, etc.) form a family with certain signal and pharmacological properties.

The search for correspondences between the amino acid sequence in a regulatory peptide and its physiological functions is the most important task of proteomics. The statistical analysis of amino acid sequences of regulatory peptides is devoted to this task. The statistical method enables to calculate the relative frequency of every amino acid residual appearance in the chain, as well as of repeated appearances of the same amino acid residual combinations. Using this method, one can reveal multiply repeating regular sequences of amino acid residuals in some proteins.

As a result of statistical analysis of 242 regulatory proteins belonging to the transcriptional factors family [Atchley et al., 2000], it was found that certain structural characteristics of polypeptides, such as positioning of amino acid residuals along the chain and tautology of certain peptide blocks, are related to the performance of their regulatory functions by these proteins: namely, with the initiation of gene transcription. This function is provided by conservative segments of the polypeptide chain with 'spiral-loop-spiral' morphology, which are specifically connected with promoter DNA sites.

Peptides regulating the functions of different physiological systems possess some common properties. On the one hand, all of them are comparatively small in size – the molecular mass of these peptides does not exceed 10 kDa – and reveal physiological activity in ultra-small doses. In many cases, the fragments of regulatory peptides reveal biological activity, modulating the activity of their predecessors: changing the diapason of activity or inhibiting it. Supposedly, the functional and tissue specificity of the effect of these peptides is connected with a variety of cellular receptors, although the possibilities of non-receptor infiltration into the cell and a direct effect of the peptide on intracellular processes cannot be disregarded at present.

On the other hand, there is a difference in the amino acid composition of peptides which were extracted from different tissues and participate in the regulation of the respective tissue-specific functions and which can be subdivided into several groups.

The amino acid composition of each peptide group can be used to compare them in terms of frequency of amino acid residual involvement. Table 13 indicates a ranked order of different amino acid residuals, which can be found in regulatory peptides belonging to different regulatory systems. Amino acid residuals possessing ionogenic lateral groups are marked by plus and minus signs, respectively. The table indicates only the 9 first ranks, because in the studied systems the amino acid residuals belonging to the first 6 ranks occupy

Table 13. Rank order of amino acid residues in regulatory peptides and proteins

Regulatory peptides and proteins group	Amino acid residues rank								
	1	2	3	4	5	6	7	8	9
Neuropeptides	L	G	A	R^+	P	S	E^-	Y	T
Gastrointestinal peptides	L	G	D^-	A	F	S	T	E^-	K^+
Immunopeptides	E^-	K^+	L	D^-	S	T	A	G	V
Atriopeptides	G	S	R^+	L	C	K^+	F	M	D^-
Aquaporin 0	L	A	G	V	F	S	R^+	T	P
Trombocyte growth factor	L	A	G	V	R^+	P	E	Q	S
Glial neurotrophic factor	L	A	R^+	D^-	S	K^+	G	V	P
Protein P-57	A	E^-	K^+	P	T	G	D^-	S	Q
Vessel endothelium growth factor	E^-	C	R^+	K^+	P	Q	F	D^-	S
Troponin T	E^-	K^+	R^+	A	L	D^-	Q	G	I

Amino acid residues with ionogenic lateral groups were marked by 'minus' and 'plus', respectively.

more than 55%, and the residuals of the first 9 ranks occupy more than 80% of all the places in the polypeptide chains. The table also includes data on several regulatory proteins.

The frequency characteristic of the amino acid composition of peptides from each group is an obvious reason for frequent repetitions of the same combinations of amino acid residuals in their chains. Less obvious is the reason for the fact that amino acid residuals, which are included in the peptide chain, follow one another with some certain predilection (selectivity). Since in the course of evolution certain selective combinations of the amino acid residuals turned out to be useful for vital functions of the organism, they were secured at the genome level.

Peptide chains include 20 different amino acid residuals. As can be seen in table 13, the amino acids differ in terms of extent of their inclusion into the composition of regular peptides. Amino acids belonging to the first four ranks form repeated blocks in the composition of their chains. Supposedly, these blocks are sources of certain molecular signals playing an informational role, because their composition differs in the peptides relevant for regulation of different physiological functions.

In the course of the study of the self-regulation mechanisms of different organisms, the signal functions of its components are of special interest. In 1968, Schmallhausen wrote: 'The links between the organism and the environment are not limited to the phenomena of substance and energy metabolism. The perception of signals, which do not play a direct role in metabolism but determine the conduct of the organism, is of great importance.'

At the molecular level, the interaction's energy to information components ratio depends on the extent of the system's organization. Presumably, among all components of live systems the highest level of molecular organization has been reached by polypeptides, which enables them to act as inductors and recipients of molecular information. Separate polypeptide fragments (regulatory peptides) may serve as the carriers of molecular information, while the DNA is the storage center for all genetic information.

The biological activity of the regulatory peptides is usually regarded from two points of view: (1) from the pharmacokinetic viewpoint, concerning the duration of their effect in the organism, and (2) in terms of pharmacology, studying the interconnection between the dose of substance (usually related to a unit of the experimental animal's weight) and the effect observed. To these ends, the scale of molar concentrations is used, enabling to compare the effect of peptides with different molecular masses. The array of active concentrations may serve as a criterion of the trophic or informational functions of the peptide.

One of the peculiarities of peptide regulation is that it is observed under conditions of very low peptide concentrations (approximately 10^{-18}–10^{-12} *M*), when the direct role of peptides in biosynthesis reactions cannot be significant [Ashmarin et al., 1992; Sazanov and Zaitsev, 1992]. As shown in previous chapters, synthetic peptides influence gene expression at a concentration of approximately 10^{-13}–10^{-14} *M*, which enables one to define their activity as a signal one.

Cytological experiments and clinical observations show that the preparations of cytomedin class also contain signal tissue-specific oligopeptides, since they, presumably, have no trophic value in the diapason of their effective concentrations (10^{-10}–10^{-12} *M*).

As a rule, the amino acid sequence in peptides, revealing a certain already identified regulatory or modulating function, can be found in the composition of one or several significantly larger proteins. Ivanov et al. [1997] put forward a hypothesis about the existence of a specific peptide pool which descends from functionally active high-molecular-weight proteins by means of tissue-specific enzyme hydrolysis of these proteins in order to obtain peptides of a certain size.

Based on the theory of endogen origin of regulatory peptides as a result of the step-by-step hydrolysis of their protein predecessors, we can suppose that it is these blocks that can reveal regulatory activity.

However, the hypotheses on the origin of regulatory oligopeptides do not define the mechanisms of their regulatory effect on cell function.

Besides an empiric search for peptide chain segments possessing informational value, a theoretical view on this problem was searched for which consisted of the statistical analysis of sequence of elements in DNA chains and polypeptides [Herzel et al., 1994; Atchley et al., 2000; Weiss et al., 2000]. This approach enables the evaluation of stable amino acid sequences (conservative blocks) in a certain protein and helps to calculate the configuration entropy of its chain and informational load of every repeated block.

The statistical view on the study of amino acid sequence differs drastically from statistical mechanics in the theory of gases. Unlike gas molecules, which can freely exchange their positions in space, the elements of the polypeptide chain cannot be rearranged at will. The fixed position of amino acid residuals along the chain complies with a certain order, and the extent of this order is entropy of the sequence.

While conducting statistical calculations, one must distinguish between precise and probabilistic information. Precise information differs from probabilistic firstly by its irreversibility: an informational 'message' and the form of its expression have a beginning, a certain sequence and an end, i.e. this order is irreversible. Written culture has been used for the long-term fixation of precise information since the humanity's early days.

Regarding intermolecular interactions between oligopeptides and nucleic acids, one can see that the sequence of the ED amino acid residuals has a structural peculiarity, which enables it to selectively interact with DNA promoter segments and to take part in the initiation of gene transcription.

The isolation of conservative amino acid blocks in structures of peptides and proteins can be used not only for evaluating their informational capacity, but also for studying the molecular mechanisms of transmission and perception of informational signals. The similarity in structure of the amino acid blocks in the composition of regulatory peptides and high-molecular-weight proteins suggests that transmission of informational signals takes place using the 'from similar to similar' principle with respect to the hypothesis of Klimov and Barashkova [1993], stating that a regulatory peptide is a molecular oscillator, and that the transmission of its informational signals is organized on the basis of the 'tuning fork-resonator' principle.

Speaking of molecular oscillators, attention must be paid to the peculiarities of their structure and to the corresponding frequency diapasons of vibrations in which they can take part.

The first characteristic feature of regulatory oligopeptides is their spatial asymmetry, which includes structural (compositional) heterogeneity, chirality, heterogeneous distribution of electrostatic charges, and a corresponding average vector of proper dipole moment.

The second peculiarity of regulatory oligopeptides is a heterogeneous distribution of hydrophilic and hydrophobic lateral groups along the chain, resulting in inhomogeneous hydration including hydration in different types of carboxyl and amino groups. This leads to the interchange of hydrate and clathrate structures of bound water, which is characteristic for this oligopeptide.

The third peculiarity of oligopeptides is the inhomogeneous distribution of the groups able to form hydrogen bonds both with the solvent and with the adjacent groups. Under certain conditions, the hydrogen bond may be accompanied by proton transfer to the proton acceptor atom and by the formation of an ion pair.

These features determine the high individual specificity of regulatory peptide interaction with the DNA double helix in the promoter segment of the gene.

Transcription of every gene is controlled by its regulatory segment, which is located near the site from which the transcription begins, in the direction from the 5′- to the 3′- end. This region is a segment of DNA double helix and regulatory peptides which are complementarily connected with it, including the transcriptional factors.

In eukaryotic cells, the transcription control is performed by a complex system of activators and transcriptional factors, including both high-molecular-weight transcriptional factors and low-molecular-weight peptides and oligopeptides. In the course of new experimental data accumulation, the quantity of proteins and peptides taking part in the control, activation and transcription of genes increases [Ryu et al., 1999]. For the purpose of unification of the designations, the further text contains one-letter symbols for amino acids, and nucleotides are marked in **bold type**.

Until recently, numerous studies were performed to prove that *Sp1* transcriptional factor consisting of 778 amino acid residuals is selectively linked to the sequence of **GC** nucleotides (**GC** boxes) on the nontranscripted segments of many genes and thus initiates their transcription. Quite recently it became obvious that not only does *Sp1* protein connects with its specific sites, but also all other proteins homologous to it, which are now united into the *Sp1* family of transcriptional factors, also do [Suske, 1999]. But the correspondences between the sequence of nucleotides at the connection site and the specific amino acid sequence of the transcriptional factor have not yet been defined.

In the past it was supposed that the amino acid residuals form hydrogen bonds with the functional groups in the DNA double helix major groove [Seeman et al., 1976]. Consequently, certain objections arose against a recognition mechanism using the 'one amino acid-one nucleotide pair' principle [Pabo and Sauer, 1992]. Really, every nucleotide pair, for example **G-C**, can by means of their functional group interact with many amino acid residuals. From 20 amino acids at least 8 (H, Y, N, Q, D, E, K, R) can interact simultaneously with two adjacent basepairs because they have sufficiently long and mobile lateral groups.

In the course of the study of repeated amino acid blocks in the structure of regulatory proteins and peptides, a certain similarity was found in these blocks among proteins and peptides controlling a certain common group of tissue-specific functions [Shataeva et al., 2002]. Most frequently, such blocks consist of 4–5 amino acid residuals, carry a certain informational load, and, presumably, can play the role of signal ligands, since they are the most evolutionarily conservative amino acid sequences in eukaryotic organisms. It is obvious that signal molecules, in particular peptides, must not be very long, or else they would quickly lose the precision of their signal function as a result of mutational changes in the genome.

We can suppose that peptide blocks and peptides containing them are activators and agonists of transcriptional factors, the primary start signal for binding the transcriptional factor to the promoter being the site-specific binding of the peptide in the DNA major groove [Shataeva et al., 2002].

As stated in previous chapters, the organism's aging is related to a decrease in mRNA concentration and, correspondingly, with retarded or terminated protein synthesis. For protein synthesis support, the transcription of RNA genes of polymerase II is required first, as well as the synthesis of this very important enzyme. More than 70 polypeptides take part in the initiation of eukaryon genes transcription, some of which are subunits of RNA polymerase II complex.

Table 14 contains the results of analysis of amino acid sequences in transcriptional factors of RNA polymerase II with lengths of 125 to 961 amino acid residuals, in which the homology of the structure is absent but in which many times repeated peptide blocks with positively and negatively charged (hydrophilic) lateral groups (ED, KE, KD, KED, EDE) are present. Adjacent to these blocks from two sides are amino acid residuals carrying hydrophobic lateral groups (A, G, I, F, P, T, V), so every tetra- or pentapeptide has its individual hydrophilic-hydrophobic sequence of lateral groups.

Analysis of nucleotide sequences in promoter segments of genes in RNA polymerase II subunits shows that these segments contain repetitions of the same nucleotide sequences by 6–10 basepairs, so their length does not exceed one loop of DNA double helix.

Molecular modelling reveals that functional groups of peptide blocks complementarily interact with nucleotides in the promoter gene segment in DNA major groove. We can suppose that polypeptides and DNA, being the two classes of biopolymers, serve as information carriers and exchange this information using binary division of informational signal.

The majority of non-transcriptable segments of other genes contain, like the peptides, numerous repetitions of same nucleotide sequences, as, for example, the representatives of the Alu family: some of them have the length of 300 basepairs, but these are predominantly repetitions of short sequences [Lodish et al., 2000]. The length of such blocks in promoter segments is usually 6–10 basepairs, i.e. it does not exceed one loop of double helix. These orderly (regular) nucleotide sequences may serve as specific sites for selective binding of peptides, including synthetic oligopeptides.

Table 14. Peptide blocks in transcription factors RNA polymerase II

Transcription factors	Blocks	Ref.	
Tf1	KEDE	AEKL	
395 aa (general	MEDQ	LKDQ	Jawhari et al. [2002]
transcription factor IIH)	IEDI		
	LDES		
Tf2	AEDA	KE^{X2}	
125aa	TEDH		Acker et al. [1993]
Zinc-finger	PKEDKEN		
	VDEL		
Tf3	AEDV	VERV	
622aa	VEDF	AREG	Ogryzko et al. [1998]
	PEDP	GDKKEP	
	PEDR		
	GDES		
Tf4	CKEDL	IEKT	
961aa	GREDQ	LEKL	Cho et al. [1999]
	NEDEG	LEKP	Mandal et al. [2002]
	ADEM	PEKKL	
	PDERDI	NKEL	
	PDREG	PDKEP	
	TDEDDH		
Tf5	QEDL	SDEQ	
349aa	IAEDET	LDEL	Zhou and Chiang [2001]
	TEEDL	QDEEDI	
	QKEDREKEQ	TKEA	
	SEDEDET	VEKEV	
	KE^{X4}	VRKRRF	
Tf6	AEDS	FDEG	
560aa	GEDG	GDEDS	Lindfors et al. [2000]
	MEDY	FEKF	
	NEDH	VEKG	
	SEEDRA	FKDKG	
	SEDDG		
	SEDDEL		
Tf7-phosphoenolpyruvate	KEDA (Livagen)	SKEF	
carboxykinase; it	SEDG	NKEW	Wilson et al. [2002]
has blocks as Tfs	VEDI	WEKEV	
	LEDQ	IEKY	
	KE^{x5}	EW	

We proposed a model of specific interaction between small peptides and recognition sites in the promoter segment of gene. This model is based on the principle of geometric and electrochemical complementarity of two right-hand molecule helices: a peptide and DNA [Khavinson et al., 2003e]. While building this model we used published data on the geometrical characteristics of the DNA double helix [Senger, 1987; Van Holde, 1988] and the peptide chain [Popov, 1997], as well as on the laws of 'polyampholyte-polyampholyte' type multifunctional interaction [Kabanov and Kabanov, 1995; Ryadnova et al., 2000].

9.1. Peculiarities of Complementary Interaction in the 'Peptide-DNA' System

Polypeptide chains consisting of *L*-isomers can have several conformations. The basic geometrical characteristics of these conformations, known from published sources and calculated on the bases of X-ray data for the polypeptide chain, are indicated in table 15. β-Structure forms of the peptide chain have the most protrusion and belong to the unfolded type (shape) of peptide conformations, which provide for the highest extent of conformational freedom of lateral groups in the peptide chain and enable them to take part in the greatest number of intermolecular interactions.

The orientation and conformation of lateral groups of amino acid residuals depend on intermolecular interactions between the peptide and components of the environment.

Regulatory (nontranscriptable) DNA segments contain sites of selective peptide binding. Nucleotide pairs in the double helix are located one over another on nearly parallel planes. One loop of undisturbed double DNA helix corresponds to 10 bp; the mean distance between the planes of basepairs makes 3.4 Å [Senger, 1987]. It is known that phosphodiether carcass chains divide the cross-section of the DNA double helix into two unequal segments. Due to this fact, the major and the minor grooves are distinguished on the molecular surface of the double helix, with widths of 22.2 and 11.8 Å, respectively, divided by two phosphodiether chains [Alberts et al., 1994; Parkinson et al., 1996].

Thus, the DNA helix and the peptide chain possess a certain metric correspondence: the length of the peptide chain in β-conformation, corresponding to one amino acid, makes 3.47 Å, while the distance between basepairs in the DNA chain is 3.4 Å. Since both participants of the interaction have a right-hand spiral orientation of basic chains, the peptide molecule can install itself into the major DNA groove without additional tension. In this case, conditions appear

Table 15. Geometrical characteristics of peptide chain [according to Alberts et al., 1994 and Popov, 1997]

Conformation name	Translational length, Å (regular sector)	Number of residues per regular sector	Distance per one amino acid residue, Å
α-Spiral	5.4	3.6	1.5
γ-Spiral	5.03	5.1	0.99
β-Spiral	13.9	4.0	3.47

for a multipoint intermolecular interaction between peptide and DNA, in which the polar and hydrophobic groups of both participants are involved. The model of special multifunctional interaction between DNA and ligands has been described and studied in detail on the example of netropsin antibiotic binding in the minor groove of the DNA double helix [Rentzeperis and Marky, 1995; Gallagher and Sharp, 1998]. The netropsin antibiotic has three hydrophobic segments and one positive charge at the end of its molecule. We extended this model on the interaction between oligopeptides and functional groups of the DNA major groove, where quite voluminous lateral groups of the peptide can be placed.

Polar interactions between the peptide and basepairs in the double helix major groove are determined by electrostatic interactions of peptide carboxyl groups with amino groups of adenine and cytosine, by hydrogen bonds of proton-carrying amino groups and carboxyl groups of the peptide with nitrogen atom ^{7}N of adenine and guanine [Harrison, 1991; Alberts et al., 1994]. Hydrophobic interactions in the system are determined by lateral groups of the peptide and methyl groups of thymin. The complementarity of electrostatic interaction between DNA and peptides was discovered already in the course of early works devoted to the study of enzyme-substrate RNAse complexes structure [Wainer et al., 1982]. Metric characteristics of these groups' location for every nucleotide pair are indicated in figure 22a. Obviously, every sequence of these pairs exposes a unique ornament of functional groups on the surface of the DNA double helix major groove.

9.2. Model of Complementary Interaction between Small Peptides and DNA

Models of site-specific DNA interaction with dipeptide KE (Vilon) and tetrapeptides KEDA (Livagen) and AEDG (Epitalon) reveal immune modulating, antioxidant and reparative activity, which leads to a decrease in tumor

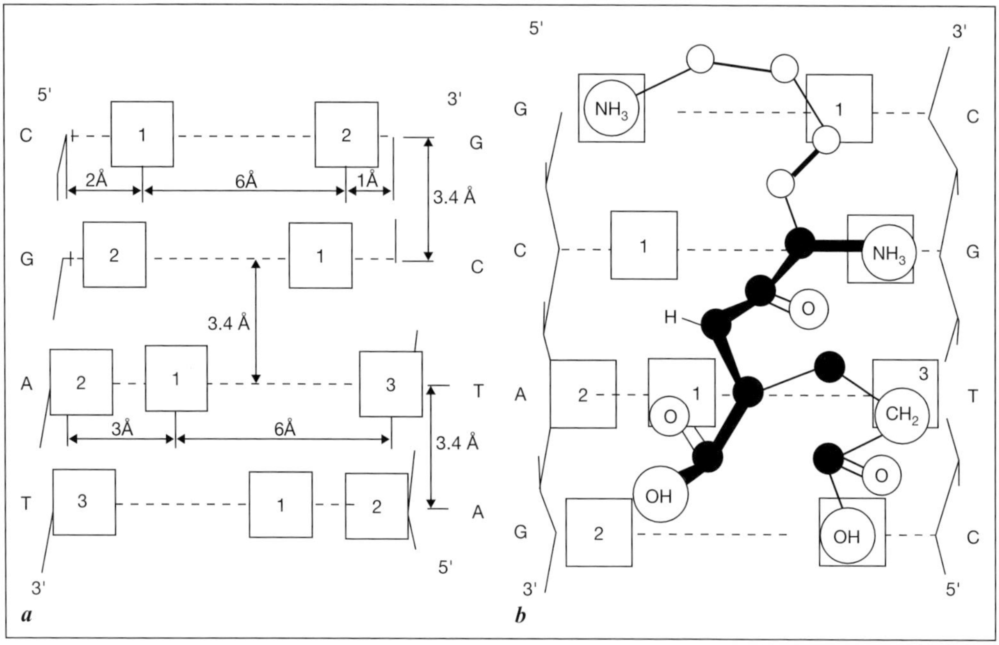

Fig. 22. Model of complementary multipoint interaction between peptides and DNA double helix. (***a***) Functional groups of DNA nucleotide pairs, exposed on the surface of double helix major groove: 1 = adenine and cytosine amino groups, which are proton donors in hydrogen bonds; 2 = ^{7}N atoms of adenine and guanine, which are proton acceptors in hydrogen bonds; 3 = methyl groups of thymine and 5-methylcytosine, participating in hydrophobic bonds with aliphatic lateral groups of peptides. (***b***) Scheme of KE dipeptide (Vilon) binding in DNA double helix major groove with **GCAG** nucleotides complementary sequence, which is repeated many times in the promoter segment of the polymerase II RNA large subunit. Bold lines and dark circles = Basic carcass of peptide chain; light circles and thin lines = lateral groups of amino acid residuals; large circles = peptide groups, participating in the interaction with DNA groups; - - - - = planes of Watson-Crick nucleotide pairs.

incidence and to an increase in the mean life span of animals [Khavinson, 2002]. It is obvious that under physiological conditions the proton-carrying amino groups of these peptides are able to take part in hydrogen bonds, while the carboxyl groups having a negative charge take part in ion-ion and polar interactions. The lateral methyl group of alanine residuals can interact with the hydrophobic DNA groups. Simple rearrangement of nucleotide pairs, the functional groups of which are shown in figure 22a, enables to define the sequence of nucleotide pairs, in which the location of their functional groups on the surface of the DNA major groove will appear complementary to the location of the peptide's lateral groups. A distinctive feature of the suggested model is the fact that the peptide in the major groove interacts simultaneously with functional groups of bases of both DNA double helix chains. Since lateral aromatic groups are not included in the three peptides selected by us, the effect of aromatic groups was not considered in the course of complementary nucleotide sequence selection.

While building the model of complementary intermolecular connections between peptides and DNA the 1.5 Å length of covalent bonds was taken into consideration, as well as the 2.5 Å length of ion-ion bond, and the 2.0–2.5 Å length of hydrogen bond. Being nonspecific, the interactions between these peptides and the phosphate groups of DNA carcass chains were not considered. The possible changes in the inclination angles of the planes of the nucleotide pairs with respect to the chain of double helix were also disregarded.

A distinct division of electrostatically charged groups is observed in the structure of the dipeptide Vilon: positive charges of lysine residuals are located in the N-end region, while the negative charges of glutamin acid residuals are located on the C-end of the molecule. These groups are unable to interact, since they are in trans-position with respect to the plane of the peptide bond, but the C-end group affects the orientation of the lysine ε-amino group. The most probable combination of nucleotide pairs, complementary to the KE sequence, corresponds with the sequence of bases **GCAG** in the direction from the 5′ to the 3′ end of DNA and is shown in figure 22b. The large lateral group length of lysine and glutamine acid residuals provides for the interaction between every one of them and two nucleotide pairs, so the binding of the peptide in the major groove appears to be an additional longitudinal stitching of the DNA double helix, the end carboxyl group of the peptide being able to interact simultaneously with two adenine groups on the **G-C** plane.

The **GCAG** nucleotide sequence shown in figure 22b is complementary to the KE dipeptide in concerning the orientation of the peptide from the N-end to the C-end in the direction from the 5′ to the 3′ end of one of the strands of the double-stranded DNA chain. In the complementary chain, the **CTGC** nucleotide sequence corresponds to it. A **GC** group is included in both sequences, **GCAG** and **CTGC**. As mentioned above, this combination of nucleotides is considered to be the general site of binding for many transcriptional factors and is contained in the nucleotide sequences of the Alu family in the form of numerous copies [Lodish et al., 2000]. Meanwhile, the phosphodiether bond in the DNA chain with a **GC** sequence is a site of attack for many endonucleases (restrictases). This is why the binding of peptide ligands to this DNA segment to a certain extent protects the **GC** sequence from enzymatic hydrolysis and thus preserves the entity of non-transcriptable genome segments.

Many years of observation have shown that aging of the organism is accompanied by the inhibition of protein synthesis. The transcription of all protein genes is performed by the enzyme of DNA-dependent RNA polymerase II after the initiation of this process by transcriptional factors [Dynlacht, 1997]. The biosynthesis of RNA polymerase proper cannot be performed without the initiation of its transcription by means of corresponding transmission factors. Basing on published data [Mita et al., 1995], the **GCAG** nucleotide sequence can be found 8 times in one chain and 7 times in the other DNA chain on the promoter segment of the large RNA polymerase II subunit. One can suppose that the experimentally discovered wide range of regulatory and modulating activity of Vilon with respect to different functions and tissues [Morozov and Khavinson, 1997; Khavinson, 2002] is determined by the fact that it binds specifically to complementary sites in the promoter segment of the gene of large RNA polymerase II subunit and serves as an agonist to the transcriptional factors initiating the transcription of this key enzyme's gene. The increase in intracellular synthesis of RNA polymerase II activates the chromatin and accelerates the synthesis of all necessary proteins.

KEDA tetrapeptide (Livagen) is an elongated variant of KE dipeptide (Vilon) and reveals more specialized tissue specificity, activating protein synthesis in hepatocytes and lymphocytes [Brodsky et al., 2001; Khavinson et al., 2002a]. Additional amino acid residuals in Livagen's structure determine the length of the nucleotide sequence with which this tetrapeptide might interact complementarily. Figure 23b shows functional groups of the **GGCATAA** sequence of bases and the KEDA tetrapeptide bound to them by multipoint interaction.

While studying the conformational features of AEDG tetrapeptide (Epitalon), it was found that its most energetically lucrative conformation in water (homogeneous environment) has a length of 15.5 Å and a maximum width of 8.5 Å. In the absence of a low-molecular-weight electrolyte, the negative charge of the C-end carboxyl group is neutralized by the proton-carrying N-end amino group, i.e. the intramolecular interaction partially cyclizes the tetrapeptide. In saline solution and under conditions of the multipoint interaction with DNA, this tetrapeptide takes on an unfolded, protruded shape.

Intermolecular interactions between the AEDG peptide and DNA take place in the form of hydrogen bonds between functional groups of both the participants. The length of the hydrogen bond in such systems usually makes 2.0–2.5 Å, i.e. exceeds the half distance between the planes of coupled double helix bases. Rearrangement of the nucleotide pairs determined their **ATTTC** sequence, which is complementary to AEDG tetrapeptide.

Figure 23 shows the location of AEDG (a) and KEDA (b) functional tetrapeptide groups, surrounded by functional group sequences complementary to them in the DNA double helix major groove.

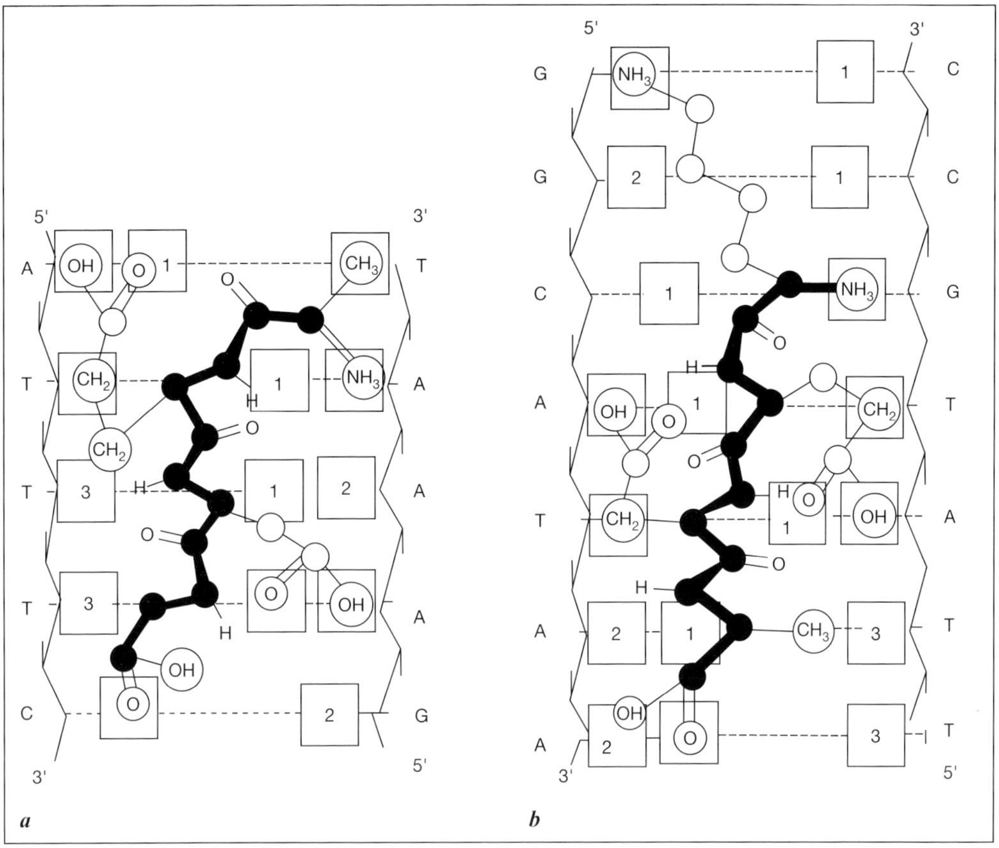

Fig. 23. Schemes of binding of peptides. (***a***) AEDG (Epitalon). (***b***) KEDA (Livagen) with complementary sequences of nucleotides in promoter segments of telomerase genes and large polymerase II RNA subunit. For further explanations of symbols, see legend to figure 22.

Noteworthy are the peculiarities of intermolecular interactions between AEDG and DNA double helix, which appear as a result of the unique amino acid residual sequence in this peptide.

Firstly, due to the mobility of the N-end amino group alanine can set up two intermolecular bonds with basepairs located on the adjacent planes (layers) of the double helix: a hydrogen bond with guanine and a hydrophobic bond with thymine, which is located in the next layer. This N-end amino acid residual can 'stitch' two DNA pairs together.

Secondly, every lateral group of ED sequence can also take part in a hydrophobic or a hydrogen bond. At the same time, the interaction is possible with nucleotide pairs **T-A** and **A-T**. A change in the sequence of amino acid residuals from ED to DE would not enable this 'stitching' of threebase pairs to be performed, because the length of the lateral group of asparagine acid is insufficient for a hydrophobic interaction with the methyl group of thymine.

Thirdly, the C-end residuum of glycine, due to the mobility of the carboxyl group, can set up the durable hydrogen bond with the **C-G** basepair.

The creation of such a complex including the peptide and a double-stranded DNA not only broadens the major groove of the double helix, but also increases its local rigidity, which may facilitate the interaction of other transcriptional factors with this certain promoter segment of the DNA.

As stated before, AEDG tetrapeptide is of special interest because of its high geroprotective efficacy [Khavinson, 2002]. Its significant effect on the increase in life span and the inhibition of the processes of malignant transformation of tissues has been shown in animal experiments. It is known that the general duration of existence of cell populations correlates with the length of chromosome telomeres and, correspondingly, with the activity of the telomerase enzyme. While comparing the information with published data [Wick et al., 1999], it was found that the **ATTTC** sequence of nucleotides, identified with the help of the complementary peptide-DNA interaction model, occurs 9 times in the structure of the promoter segment of the telomerase protein component gene (3,729 bp on the segment since the beginning of transcription). Presumably, the complementary interaction between AEDG oligopeptide and **ATTTC** blocks leads to the reactivation of telomerase promoter in somatic cells, which initiates intracellular telomerase synthesis, elongation of the telomeres, as well as an increase in the proliferative potential of tissues and thus can exert a significant influence on life span.

On the promoter segment of the RNA polymerase II large subunit gene, the **ATTTC** sequence also occurs 3 times.

It is noteworthy that the similar metrics of the location of the functional groups on the surface of the double helix major groove for **A-T** and **G-C** nucleotide pairs (fig. 23a), the high mobility of the functional groups of end amino acid residuals of alanine and glycine, as well as the variability of the inclination angle of the plane of the nucleotide pairs with respect to the axis of the DNA double helix, enable the AEDG tetrapeptide to complementarily bind with the **GTTTC**, **CTTTC** and **ATTTG** sequences, although the durability of these bonds will be lower than with **ATTTC**.

In the past, the **ATTTGCAT** block was reported to be one of the binding sites for the transcriptional factors of RNA polymerase II [Mitchell and Tijan, 1989]. Table 16 includes these data, as well as the data on some binding sites of transcriptional factors in promoter segments of the gene of the T cell activation nuclear factor, *NFATc*. We can suppose that AEDG synthetic tetrapeptide reveals its geroprotective and tissue-specific effect as an agonist of transcriptional

Table 16. Amino acid sequences of transcriptional factors and peptides, and their corresponding nucleotide sequences – sites of recognition in the DNA chain

Transcriptional factor chain region		Site of binding
Legend	Sequence	
Oct-2* homeodomain	RRRKKRTSIE TNVRFALEKS	Oct-1 ATGCAAAT
[Mitchell, Tijan, 1989]	FLANQKPTSE	TACGTTTA
	EILLIAEQLH MEKEVIRVWF	Oct-2 ATTTGCAT
	CNRRQKEKRIN	TAAACGTA
NFATc-DNA binding	PKSHHRAHYETEGSRGAVKA[40]	GGAAAA
domain*	SAGGHP	CCT T T T
[Yaseen et al., 1993;	site: RAHYETEGSR	GGAGGAAAAACTGTTTC
Wolfe et al., 1997]		ATACAGAAGGCGT
Polymerase II RNA gene	KE	$GCAG_{x15}$
transcription activator	AEDG	$ATTTC_{x3}$
	KEDA	GGCATAA
Telomerase gene		
transcription activator	AEDG	$ATTTC_{x9}$
Retinal gene F379	AEDG	GTTTC; $CTTTC_{x2}$
transcription activator	KE	$GCAG_{x7}$; $CTGC_{x4}$

*Participate in initiating of RNA polymerase II synthesis.
Lower index shows the number of repeats of this particular nucleotide block in the regular gene sector.

factors, interacting with nucleotide blocks in the promoter segments of vitally important genes.

Analysis of the structure of the gene, which is specific for retina, is of certain interest [Mah et al., 2001]. A comparatively short promoter segment of this gene (346 bp) contains **GTTTC** and **ATTTA** blocks, as well as repetitions of the **CTTTC** nucleotide sequence, i.e. the sites of possible binding of AEDG tetrapeptide, and a 7-fold repetition of **GCAG** block in the major groove – the possible sites of Vilon binding.

It has already been proved that the polypeptide retinal preparation (Retinalamin) in case of its effect on early gastrula ectoderm (polypotent tissue) at a concentration of 10 ng/ml invoked neural induction, including brain, retina and pigment epithelium [Khavinson et al., 2002b]. In other words, it was proven experimentally that the peptides are able not only to stimulate gene transcription and protein synthesis in differentiated tissue, but also to take part in cell cycle control. Supposedly, this very feature underlies the anticarcinogenic effect of AEDG tetrapeptide [Anisimov et al., 2002].

Table 16, apart from the published data on amino acid sequences of natural transcription factors and on the corresponding specific nucleotide sequences (binding sites), contains the amino acid sequences of synthetic peptides and the complementary sequences of nucleotide pairs, which we identified using the proposed model.

The proposed model is not an attempt to have a precise description of the complementary interaction in the peptide-DNA system, since the conformations of the interaction of the participants used were calculated for a homogeneous environment, i.e. are true on condition that interactions with other molecules are absent. Besides, we did not take the effect of the aqueous surrounding in the region of the intermolecular DNA interactions into consideration. Nevertheless, the proposed model is worthy of experimental study and specifications with the use of isotopically marked peptide and DNA links.

9.3. Informational Interaction in the 'Peptide-DNA' System

Complementary interaction between peptide chain and DNA double helix not only broadens the major groove of the double helix, but also increases the local rigidity of the chains of these biopolymers. A corresponding decrease in conformational entropy of both molecules can be regarded as an informational signal. However, for quantitative study of the signal appearing as a result of a complementary interaction between peptide and DNA, a statistical evaluation of the informational load is required regarding each of interaction's participants. In other words, we need to estimate how many bits of information one structural symbol (letter) in the repeated sequence of symbols in peptide and nucleotide chains may contain.

The quantity of repeated amino acid blocks in the structure of proteins and peptides was estimated in the previous study, as well as an evaluation of the informational load according to Shannon, which corresponds to every amino acid residuum in the sequence [Shataeva et al., 2003]. The calculation according to the original program [Chernova A.A. http://www.maths.ox.ac.uk./~chernova/codes/belok.htm] showed that the most frequently repeated blocks in the composition of polypeptide chains consist of 4–5 amino acid residuals and carry a certain informational load, which in some cases reaches 8 bits per amino acid residual in the block. The repeated blocks in peptide chains, carrying informational load, are the most evolutionarily conservative amino acid sequences in eukaryotic organisms. Obviously, the signal molecules, in particular peptides, must not be very long or else they would soon lose the precision of their signal function as a result of mutational changes.

Similarly, the evaluation of statistical information was conducted on nucleotide sequences in the promoter segments of RNA polymerase II genes, as well as on the large telomerase subunit, preproinsulin and proopiomelanocortin. The calculation showed that non-coding segments of these genes contain repeated nucleotide blocks, most frequently consisting of 7 parts, for which the value of the informational load reaches 6.4 bits per basepair. Since amino acid and nucleotide 'alphabets' contain different numbers of symbols (20 and 4 symbols, respectively), translation of information from one language to another without losses and distortions may occur if the information donor and acceptor have a comparable informational capacity. For example, informational signal transition from AEDG tetrapeptide to DNA requires repeated nucleotide blocks, consisting of 5–7 nucleotide pairs.

Thus, the molecular model of complementary interaction between the small peptides and the DNA double helix is supplemented by the concept of dosed information transition in the case of such contact. Short blocks (oligomers), being preserved during millions of years of evolutionary development in the structure of high molecular biopolymers – proteins and nucleic acids, are information carriers.

10

From Anti-Aging Peptide Regulation to Pharmacogenomics

Nowadays, the fact that there are genes which affect life span and longevity is undoubted [Kirkwood, 2003]. Long-term studies devoted to the molecular-genetic mechanisms of aging enabled the way in which genes affect the life span and aging of certain organisms to be revealed. There are numerous examples of how the differences in gene expression or activity lead to differences in the life span of related organisms. There are also changes at the genome level which are connected with aging. Chromosomal disorders in somatic cells are an illustrious example of this fact. However, for the majority of species these genetic mechanisms remain unknown. Due to insufficient data in our possession, an impression arises that the genetic mechanisms of aging are much more complex and reveal greater specific variability than the genetic mechanisms of development. However, so far there was no need to specify some principally new biological mechanisms, which would differ from those taking part in the process of organism development, for the explanation of aging. Moreover, there is an impressive number of examples where biochemical and cellular functions remain unchanged throughout life. Studies in development biology showed that the majority of human somatic cells possess a certain proliferative potential and the ability to differentiate throughout life. Therefore, in the case of an age-related decrease in these functions, the possibility arises to influence aging processes by means of the effect on gene expression [Frolkis, 1970; Finch, 1990; Khavinson, 2002; Anisimov, 2003].

The study of aging mechanisms and the development of a theoretical prerequisite for the creation of effective bioregulators enabling to reach the specific limit of life span and to retain basic physiological functions is one of the most burning issues of modern biogerontology. Studies devoted to this problem focus on the role of peptides in the prevention of aging.

Both experimental and clinical studies revealed the effects of peptide bioregulators to resist age-related alterations. As published data indicate [Khavinson, 2002; Anisimov, 2003], the expression of genes, the products of which take part in inflammatory reactions, increases in the course of aging, while the expression of genes participating in the energy supply of cells, in their differentiated functions, as well as in cell proliferation control, decreases. It is noteworthy that inflammatory reactions significantly contribute to the pathogenesis of the most radical age-related disorders in the human organism, e.g, atherosclerosis and Alzheimer's disease. The predisposition to inflammatory processes is a consequence of immune imbalance, related to prevalent weakening of T-dependent immune reactions. This weakens the organism's resistance against many infections and cancer diseases, the incidence of which grows in the course of aging. Another factor of tumor incidence increase consists of the weakening of cell proliferation and differentiation control, which not only contributes to tumor growth, but also disturbs tissue homeostasis, as well as regeneration and normal functions of organs and tissues. The weakening of the latter is also caused by disorders in the mitochondrial energy production mechanisms. The disturbances in mitochondrial functions lead to the growing production of active oxygen forms, which serve as factors providing for tissue damage and the development of cancer, atherosclerosis and age-related nervous disorders. Among age-related changes in differentiated tissues, of major importance are the processes going on in the neuroendocrine organs, since they affect the functions of other systems under corresponding hormonal control. Keeping this in mind, we can estimate the results of the study of the molecular-genetic mechanisms of the effect of peptides which is aimed at the prevention of aging.

It is known that aging involves the expression of many changes in the genes, especially those which regulate the production of regulatory peptides. A regulatory imbalance, resulting from such changes, may be compensated or corrected by peptide bioregulators, which normalize the functions of certain organs and tissues and exert positive system effects, including the enhanced energy supply of tissues, reduced generation of active oxygen forms, and the prevention or correction of age-related disturbances in immune, endocrine and other functions. On the whole, this constitutes the geroprotective effect demonstrated both by experiments and in clinical practice.

The present study confirmed the assumption that peptide bioregulators can control gene expression and the processes of protein synthesis in the cells. In vivo administration of Vilon and Epitalon to mice appeared to change the profile of gene expression in the myocardium. The alterations in gene expression are the most probable reason for the peptides' physiologic effects, such as the restoration of melatonin secretion rhythm in case of Epitalon administration to old Rhesus macaques, or the prolongation of the period of eye retina being functionally and morphologically intact in the case of Epitalon administration to rats with hereditary pigment dystrophy of the retina [Khavinson, 2002]. It is the effect of Vilon and Epitalon on the expression of IL-2 and *c-fos* genes that probably provides for the immune modulating, oncologically modifying and stress protective properties of these peptides. Data received from the study of the effect of Vilon and Epitalon on carcinogenesis in transgenic mice, confirm this idea and suggest the effect of the studied peptides on the expression of the HER-2/neu breast cancer gene. Taking into consideration the interconnection of the processes of free radical oxidation and aging, the antioxidant and antimutagenic effect of Vilon and Epitalon is of great importance. For the first time the mechanisms of the small peptides geroprotective effect, connected with chromatin activation in blood lymphocytes of senile patients were revealed. This opens new possibilities for the prevention of accelerated aging, as well as for age-related pathology correction. The effects of Epitalon, reconnected with the increase in telomerase enzymatic activity, telomere elongation in different cells and prolonged life cycle of human diploid cells due to the overcoming of Hayflick's limit not only constitute the greatest achievement of present-day biogerontology, but also confirm good prospects for the development of peptide geroprotective means.

The study on the impact of peptides on the expression of different genes and on DNA synthesis in different experimental models permitted a concept to be set forth that small peptides are activators and agonists of transcriptional factors, the primary start signal for the binding of transcriptional factor and promoter being the site-specific binding of the peptide in the DNA major groove.

We assume that the geroprotective effect of peptide regulators is associated with the regulation of gene expression and biosynthesis process, disturbed in the course of aging. This is even more important if we take into consideration the determining role of genetic apparatus in the mechanisms of individual development and the occurrence of many diseases.

Nowadays, it is well known that the efficacy of the action of a medicinal substance depends on many variables such as age, functional status of organs, accompanying therapy, chemical nature of medications and their interaction. The contribution of genetic factors to reaction variability is 20–95%. Presently, a great deal of examples have been accumulated showing that the differences in the response to drug therapy are determined by the variants of nucleotide sequence in genes, coding the medication's metabolism enzymes, medication-carrying molecules, as well as receptors interacting with medications. In most cases, a therapeutic effect is determined by the interaction of several gene products which affect the preparation's pharmacokinetics and pharmacodynamics. The future of pharmacogenomics is in the clinical application of molecular methods aimed at the individual selection of a medicinal preparation and its dose for the patient. Genotyping will play an important role in clinical practice if it is interpreted in compliance with the diagnosis and used correctly in therapy selection. The results of pharmacogenomic studies must be put into the foundation of molecular methods of therapy selection [Evans and McLeod, 2003].

The results of our studies, especially those with DNA microarrays employment, point out that small peptides are not only the means of individual therapy revealing good prospects, but also preparations, the effect of which is aimed at supporting lifelong genetic stability.

We admit that the presented data on molecular-genetic mechanisms of the effect of small peptides are still fragmentary, but confirm modern tendencies towards the development of proteomics and pharmacogenomics and set forth new approaches to the prevention of age-related pathology development.

11

Conclusion

The increase of active human life span is one of the principal tasks of gerontology and modern preventive medicine at large. Current anti-aging medicine contributes to our knowledge in the mechanisms of aging and reveals new prospects for the invention of geroprotective means which are physiologically adequate for humans. Comprehensive study of the mechanisms of peptide effect gives every evidence that this class of compound meets the requirements most fully. Classical data reflecting the normalizing effect of peptides on age-related hormonal-metabolic and immune changes are presently supplemented with the results of a genome study of peptide action.

The study of molecular-genetic mechanisms of the effect of peptides enables a new concept to be set forth, which most fully reflects the evolutionary-biological role of peptides in the organism. Peptides and DNA are two classes of biopolymers, which carry information and exchange it in the process of the initiation of gene transcription. The interaction of peptides and DNA contributes to the restoration of gene structure and expression, thus providing genetic stability and normalization of age-related metabolic disturbances. All this serves to the prevention of age-related pathology development, as well as the increase in life span up to the specific limit.

References

Acker J, Wintzerith M, Vigneron M, Kedinger C: Structure of the gene encoding the 14.5 kDa subunit of human RNA polymerase II. Nucleic Acids Res 1993;21:5345–5350.

Alberts B, Bray D, Lewis J, et al: Molecular Biology of the Cell, ed 3. New York/London, Garland Publishing, 1994.

Alimova IN, Bashurin DA, Popovich IG, et al: Effect of Epitalon and Vilon treatment on mammary carcinogenesis in transgenic erbB-2/NEU mice. Vopr Onkol 2002;48:57–60.

Andrews GR: The priority of basic research on aging vulnerability in a comprehensive research agenda on aging for the 21st century. Novartis Found Symp 2001;235:4–9; discussion 9–10.

Anisimov SV, Boheler KR, Khavinson VKh, Anisimov VN: Studies of the effects of Vilon and Epitalon on gene expression in mouse heart using DNA-microarray technology. Bull Exp Biol Med 2002;133:293–299.

Anisimov VN: Molecular and Physiological Mechanisms of Aging. St. Petersburg, Nauka, 2003.

Anisimov VN: The physiological functions of the epiphysis (the gerontological aspect). Ross Fiziol Zh Im I M Sechenova 1997;83:1–13.

Anisimov VN, Alimova IN: The use of mutagenic and transgenic mice for the study of aging mechanisms and age pathology. Adv Gerontol 2001;7:72–94.

Anisimov VN, Khavinson VKh, Mikhalski AI, Yashin AI: Effect of synthetic thymic and pineal peptides on biomarkers of aging, survival and spontaneous tumor incidence in female CBA mice. Mech Age Dev 2001;122:41–68.

Anisimov VN, Khavinson VKh, Popovich IG, Zabezhinski MA: Inhibitory effect of peptide epitalon in colon carcinogenesis induced by 1,2-dimethylhydrazine in rats. Cancer Lett 2002;183:1–8.

Anisimov VN, Khavinson VKh, Provinciali M, et al: Inhibitory effect of the peptide epitalon on the development of spontaneous mammary tumors in Her-2/NEU transgenic mice. Int J Cancer 2002;101:7–10.

Anochin KV, Ryabinin AE. Expressions of c-fos and c-jun genes in the neocortex and hippocampus of mice after passive avoidance learning. Int J Memory 1993;1:67–70.

Arking R: The Biology of Aging, ed 2. Sunderland, Sinauer, 1998.

Ashmarin IP, Lelekova TV, Sanzhieva LTs: The effectiveness of ultralow doses and concentrations of biologically active compounds. Izv Akad Nauk SSSR Biol 1992;4:531–536.

Atchley WR, Wollenberg KR, Fitch WM, et al: Correlation among amino acid sites in bHLH protein domains: An information theoretic analysis. Mol Biol Evol 2000;17: 164–178.

Arushanyan EB, Arushanyan LG, El'kebyan KS: The place of pineal adrenocorticotrope relationship in the correction regulation of behaviour. Uspekhi Physiol Sci 1993;24:12–23.

Ashmarin IP, Koroleva SV: Regulatory Peptides. Bioinformative Approach to the Investigation of Functional Continuum. Advances in Functional Neurochemistry. St. Petersburg, St. Petersburg State University Press, 2003.

Bach I: The LIM domain: Regulation by association. Mech Dev 2000;91:5–17.
Badiu C, Coculescu M, Moller M: Arginine vasotocin mRNA revealed by in situ hybridization in bovine pineal gland cells. Cell Tissue Res 1999;295:225–229.
Bartsch H, Buchberger A, Franz H, et al: Effect of melatonin and pineal extracts on human ovarian and mammary tumour cells in a chemosensitivity assay. Life Sci 2000;67: 2953–2960.
Belokrylov GA, Popova OYa, Derevnina ON, et al: Immuno-, phagocytosis modulating and antitoxic properties of amino acids and peptide preparations. Drug Dev Ind Pharm 1998;24:115–127.
Benson B, Machen N, Dunn AM, Wise ME: Chronic lateral ventricle infusion of a pineal gland–derived decapeptide alters pulsatile secretion of LH in rats. Life Sci 1996;58: 1083–1090.
Botchkov NP: Genetics for the medicine of the 21st century. Vestnik Russ Military Medl Acad 1999;1:44–47.
Bodey B, Kaiser HE: Development of Hassall's bodies of the thymus in humans and other vertebrates (especially mammals) under physiological and pathological conditions: immunocytochemical, electronmicroscopic and in vitro observations. In Vivo 1997;11: 61–85.
Boldyrev AA: Carnosin. Biological Significance and Possibilities of Medical Application. Moscow, Moscow State University Press, 1998.
Bonavera JJ, Swerdloff RS, Sinha Hakim AP, et al: Aging results in attenuated gonadotropin releasing hormone-luteinizing hormone axis responsiveness to glutamate receptor agonist N-methyl-*D*-aspartate. J Endocrinol 1998;10:93–99.
Bredas J-L (ed): Conjugated Oligomers, Polymers, and Dendrimers: From Polyacetilene to DNA. Proc 4th Francqui Colloquium. Paris, De Berg University, 1998.
Brodsky VYa, Khavinson VKh, Zolytarev Yu A, et al: Rhythm of protein synthesis in cultures of hepatocytes from rats of different ages. Norm and effect of peptide Livagen. Izv AN Ser Biol 2001;5: 517–521.
Bullitt E: Expression of c-fos-like protein as a marker for neuronal activity following noxious stimulation in the rat. J Comp Neurol 1990;296:517–530.
Campbell IL, Stalder AK, Akwa Y, et al: Transgenic models to study the actions of cytokines in the central nervous system. Neuroimmunomodulation 1998;5:126–135.
Cao SX, Dhahbi JM, Mote PL, Spindler SR: Genomic profiling of short- and long-term caloric restriction effects in the liver of aging mice. Proc Natl Acad Sci USA 2001;98:10630–10635.
Chakravarti B, Abraham GN: Aging and T-cell-mediated immunity. Mech Aging Dev 1999;108: 183–206.
Chazov EI, Isachenkov VA: Pineal gland: The Place and the Role in the System of Neuroendocrine Regulation. Moscow, Nauka, 1974.
Cheido MA, Idova GV, Pansuevich OS: The peptide modulation of immune reactions. Eksp Klin Farmakol 1993;56:51.
Chen X, Herbert J: Regional changes in c-fos expression in the basal forebrain and brainstem during adaptation to repeated stress: Correlations with cardiovascular, hypothermic and endocrine responses. Neuroscience 1995;64:675–685.
Chernova AA: http://www.maths.ox.ac.uk./~chernova/codes/belok.htm
Chipens GI, Freidlin IS, Skliarova SN: Origin and evolution of peptide-protein bioregulators. Zh Evol Biokhim Fiziol 1987;23:361–372.
Cho H, Kim TK, Mancebo H, et al: A protein phosphatase functions to recycle RNA polymerase II. Genes Dev 1999;13:1540–1552.
Darcissac EC, Bahr GM, Pouillart PR, et al: Selective potentiation of cytokine expression in human whole blood by murabutide, a muramyl dipeptide analogue. Cytokine 1996;8: 658–666.
Dawson D, van den Heuvel CJ: Integrating the actions of melatonin on human physiology. Ann Med 1998;30:95–102.
Debeljuk L, Lasaga M: Modulation of the hypothalamo–pituitary–gonadal axis and the pineal gland by neurokinin A, neuropeptide K and neuropeptide gamma. Peptides 1999;20:285–299.
Derenzini M, Sirri V, Pession A, et al: Quantitative changes of the two major AgNOR proteins, nucleolin and protein B23, related to stimulation of rDNA transcription. Exp Cell Res 1995;219: 276–282.
Dilman VM: Ontogenetic model of aging and disease formation and mechanisms of natural selection. J Theor Biol 1986;118:73–81.

Dilman VM: Aging, rate of aging and cancer. A search for preventive treatment. Ann NY Acad Sci 1994; 719:454–455.
Duncan MJ, Herron JM, Hill SA: Aging selectively suppresses vasoactive intestinal peptide messenger RNA expression in the suprachiasmatic nucleus of the Syrian hamster. Brain Res Mol Brain Res 2001;87:196–203.
Dynlacht BD: Regulation of transcription by protein, that control the cell cycle. Nature 1997;389: 149–152.
Elmore LW, Holt SE: Telomerase and telomere stability: A new class of tumor suppressor? Mol Carcinog 2000;28:1–4.
Erlwanger RH, Unmack MA, Grondahl ML, et al: Effect of age on vasoactive intestinal polypeptide-induced short-circuit current in porcine jejunum. Comp Biochem Physiol Mol Integr Physiol 1999;124:29–33.
Evans WE, McLeod HL: Pharmacogenomics drug disposition, drug targets, and side effects. N Engl J Med 2003;348:538–549.
Finch C: Longevity, senescence and the genome. Univ of Chicago Press, Illinois, 1990.
Frolkis VV: On the regulatory mechanism of molecular-genetic alterations during aging. Exp Geront 1970;5:37–47.
Gallagher K, Sharp K: Electrostatic contributions to heat capacity changes of DNA-ligand binding. Biophys J 1998;75:769–776.
Garrity PA, Chen D, Rothenberg EV, Wold BJ: Interleukin-2 transcription is regulated in vivo at the level of coordinate binding of both constitutive and regulated factors. Mol Cell Biol 1994;14: 2159–2169.
Goldstein R: Arginine-vasotocin (AVT) – a pineal hormone in mammals. Rom J Endocrinol 1992;30: 21–44.
Golubev AG: Catecholamines, steroids and aging of the nervous and endocrine systems. Uspekhi Sovremennoi Biol 1989;6:64–75.
Grishina TV, Mulberg AA, Golovko OI, et al: Physical-chemical characteristics of nuclear proteins isolated from brain tissue and spleen cells regulating IL-2 gene expression. Neurochemistry 1996;13:90–97.
Grokhovsky SL, Surovaya AN, Zhuze AL, Gursky GV: Interaction of synthetic peptide containing a part of DNA-binding domain of v-Jun transcription activator with DNA. Mol Biol 1994a;28:1128–1136.
Grokhovsky SL, Surovaya AN, Zhuze AL, Gursky GV: Synthesis and binding to DNA of the two peptides, modeling the DNA-binding domain of transcription activator v-Jun. Mol Biol 1994b; 28:1137–1148.
Gutman GM: Global dialogue, global solutions. IAG Newsletter 2002;17:1–2.
Gutterman JU: Biologic therapy of human cancer. Cancer 1992;70:906–908.
Harman D: Aging and disease: Extending functional life span. Ann NY Acad Sci 1996;786:321–336.
Harrison SC: A structural taxonomy of DNA-binding domains. Nature 1991;353:715–719.
Hayflick L: The future of aging. Nature 2000;408:267–269.
Heliot L, Mongelard F, Klein C, et al: Nonrandom distribution of metaphase AgNOR staining patterns on human acrocentric chromosomes. J Histochem Cytochem 2000;48:13–20.
Hemminki A. The molecular basis and clinical aspects of Peutz–Jeghers syndrome. Cell Mol Life Sci 1999;55:735–750.
Herzel H, Ebeling W, Schnutt AO: Entropies of biosequences: The role of repeats. Phys Rev E Stat Phys Plasmas Fluids Relat Interdiscip Topics 1994,50:5061–5071.
Hipkiss AR, Brownson C, Carrier MJ: Carnosine, the anti-aging, anti-oxidant dipeptide, may react with protein carbonyl groups. Mech Aging Dev 2001;122:1431–1445.
Hiyama E, Hiyama K, Yokoyama T, et al: Correlating telomerase activity levels with human neuroblastoma outcomes. Nat Med 1995;1:249–255.
Honkaniemi J, Kainu T, Ceccatelli S, et al: Fos and Jun rat central amygdaloid nucleus after stress. Mol Neurosci 1992;3:849–852.
Hunt SP, Pint A, Evan G: Induction of c-fos like protein in spinal cord neurons following sensory stimulation. Nature 1987;328:632–663.
Hwang DM, Dempsey AA, Wang RX, et al: A genome-based resource for molecular cardiovascular medicine: Toward a compendium of cardiovascular genes. Circulation 1997;96:4146–4203.

Ivanov SD, Kovan'ko EG, Popovich IG, Zabezhinskii MA: Assessment of genotoxicity and late effects of radiation and chemical exposures. Radiats Biol Radioecol 1999;39: 418–424.

Ivanov VT, Karelin AA, Philippova MM, et al: Hemoglobin as a source of endogenous bioactive peptides: The concept of tissue-specific peptide pool. Biopolymers 1997;43:171–188.

Jawhari A, Laine JP, Dubaele S, et al: p52 mediates XPB function within the transcription / repair factor TF II H. J Biol Chem 2002;277:31761–31767.

Kabanov AV, Kabanov VA: DNA Complexes with polycations for the delivery of genetic material into cells. Bioconjugate Chem 1995;6:7–20.

Kaskow JW, Regmi A, Mulchahey JJ, et al: Changes in brain corticotropin-releasing factor messenger RNA expression in aged Fischer 344 rats. Brain Res 1999;822:228–230.

Kayo T, Allison DB, Weindruch R, Prolla TA: Influences of aging and caloric restriction on the transcriptional profile of skeletal muscle from rhesus monkeys. Proc Natl Acad Sci USA 2001;98: 5093–5098.

Kazakova TB, Burov SV, Golovko OI, et al: Biological activity of analogs of peptide hormone luliberin in the regulation of immune response of T lymphocytes. Bull Eksp Biol Med 1996;122:334–337.

Khavinson VKh: Peptides and aging. Neuro-Endocrinol Lett 2002;23(suppl 3):special issue.

Khavinson VKh: Tetrapeptide revealing geroprotective effect, pharmacological substance on its basis, and the method of its application. US 6,727,227, 2004.

Khavinson VKh, Barinov VA, Arutjunyan AV, Malinin VV: Free Radical Oxidation and Aging. St. Petersburg, Nauka, 2003a.

Khavinson VKh, Bondarev IE, Butyugov AA: Epitalon peptide induces telomerase activity and telomere elongation in human somatic cells. Bull Exp Biol Med 2003b;135:590–592.

Khavinson VKh, Bondarev IE, Butyugov AA, Smirnova TD: The peptide promotes the overcoming of the limit of somatic human cells division. Bull Exp Biol Med 2004a;137: 573–577.

Khavinson VKh, Lezhava TA, Malinin VV: Effects of short peptides on lymphocyte chromatin in senile subjects. Bull Exp Biol Med 2004b;137:78–81.

Khavinson VKh, Lezhava TA, Monaselidze JG, et al: Effects of Livagen peptide on chromatin activation in lymphocytes from old people. Bull Exp Biol Med 2002a;134:389–392.

Khavinson VKh, Malinin VV: Mechanisms underlying geroprotective effects of peptides. Bull Exp Biol Med 2002;133:1–5.

Khavinson VKh, Malinin VV, Trofimova SV, Zemchikhina VN: Inductive activity of retinal peptides. Bull Exp Biol Med 2002b;134:482–484.

Khavinson VKh, Zhukov VV: Thymus peptides and mechanisms of immunomodulation. Uspekhi Sovremennoi Biol 1992;112:554–570.

Khavinson VKh, Morozov VG: Peptides of Pineal Gland and Thymus in Aging Regulation. St. Petersburg, Foliant, 2001.

Khavinson V, Morozov V: Peptides of pineal gland and thymus prolong human life. Neuro-Endocrinol Lett 2003;24:233–240.

Khavinson VKh, Morozov VG, Malinin VV, Grigoriev EI: Tetrapeptide stimulating functional activity of neurons, pharmacological agent based thereon and method of use thereof. EP 1 224 212 B1, 2003c.

Khavinson VKh, Morozov VG, Malinin VV, Sery SV: Use of a dipeptide for stimulating repair processes. US 6,642,201 B1, 2003d.

Khavinson VKh, Shataeva LK, Bondarev IE: A model of interaction between regulatory peptides and DNA double helix. Uspekhi Sovremennoi Biol 2003e;123:467–474.

Khelimsky AM: Epiphysis (Pineal Gland). Moscow, Meditsina, 1969.

Khirokava K: Thymus and aging, in Makinodan T, Yunis E (eds): Immunology and Aging. Moscow, Mir, 1980.

Kida S, Pantazis A, Weller RO: CSF drains directly from the subarachnoid space into nasal lympatics in the rat: Anatomy, histology and immunological significance. Neuropathol Appl Neurobiol 1993; 19:480–488.

Kirkwood TB: Genes that shape the course of aging. Trends Endocrinol Metab 2003;14: 345–347.

Klimov PK, Barashkova GM: Endogenous peptides as a single system of regulatory substances. Fiziol Zh Im I M Sechenova 1993;79:80–87.

Korkushko OV, Khavinson VKh, Butenko GM, Shatilo VB: Peptide Preparations of Thymus and Pineal Gland in Prevention of Accelerated Aging. St. Petersburg, Nauka, 2002.

Korneva EA, Potin VV: Effect of destruction of the posterior hypothalamic nuclei on the thyrotropic and exophthalmic activity of the rabbit hypophysis. Fiziol Zh SSSR Im I M Sechenova 1970;56: 159–164.

Korneva EA, Shhinek EK: Hormones and Immune System. Leningrad, Nauka, 1988.

Korochkin LI, Mikhailov AT: Introduction in Neurogenetics. Moscow, Nauka, 2000.

Kovacs KJ: C-fos as a transcription factor: A stressful (re)view from a functional map. Neurochem Int 1998;33:287–297.

Kovaks KJ, Sawchenko PE: Glucocorticoid negative feedback is exerted selectively on vasopressin gene transcription in parvocellular neurosecretory neurons. Soc Neurosci Abstr 1997;27:797–798.

Kurschner C, Mermelstein PG, Holden WT, Surmeier DJ: CIPP, a novel multivalent PDZ domain protein, selectively interacts with Kir4.0 family members, NMDA receptor subunits, neurexins, and neuroligins. Mol Cell Neurosci 1998;11:161–172.

Kuzis K, Coffin JD, Eckenstein FP: Time course and age dependence of motor neuron death following facial nerve crush injury: Role of fibroblast growth factor. Exp Neurol 1999;157:77–87.

Kuznik BI, Morozov VG, Khavinson VKh: Cytomedins: 25 Years of Experimental and Clinical Study. St. Petersburg, Nauka, 1998.

Kuznik BI, Pinelis IS, Khavinson VKh: Application of Peptide Bioregulators in Stomatology. St. Petersburg, Aesculapius, 1999.

Lear JD, Schneider JP, Kenker PK: Electrostatic effects on ion selectivity and rectification in designed ion channel peptides. J Am Chem Soc 1997;119:3212–3217.

Leo A, Hansch C, Elkins D: Partition coefficients and their uses. Chem Rev 1971;6:525–616.

Levine JE, Prystowsky MB: Polypeptide growth factors in the nucleus: A review of function and translocation. Neuroimmunomodulation 1995;2:290–298.

Lezhava TA: Chromosomes in Very Senile Age: 80 Years and Over. Moscow, Nauka, 1999.

Lezhava TA: Chromosome and aging: Genetic conception of aging. Biogerontology 2001;2:253–260.

Lindfors K, Halttunen T, Huotari P, et al: Identification of novel transcription factor-like gene from human intestinal cells. Biochem Biophys Res Commun 2000;276:660–666.

Lodish H, Berk A, Zipursky SL, et al: Molecular Cell Biology, ed 4. London, Freeman, 2000.

Lundgren M, Chow C, Sabbattini P, et al: Transcription factor dosage affects changes in higher order chromatin structure associated with activation of a heterochromatic gene. Cell 2000;103:733–743.

Mah N, Stoehr H, Schulz HL et al: Identification of a novel retina-specific gene located in a subtelomeric region with polymorphic distribution among multiple human chromosomes. BBA 2001; 1522:167–174.

Mandal SS, Cho H, Kim S, et al: FCP1, a phosphatase specific for the heptapeptide repeat of the largest subunits of RNA polymerase II, stimulates transcription elongation. Mol Cell Biol 2002;22: 7543–7552.

Marin F, Boya J, Calvo JL, et al: Immunocytochemical localization of basic fibroblast growth factor in the rat pineal gland. J Pineal Res 1994;16:44–49.

Marsili V, Mancinelli L, Calzuola I, Gianfranceschi GL: Small acidic peptides and peptidomimetic molecules: cell growth and differentiation. Riv Biol 2000;93:175–181.

Mazzone SB, Geraghty DP: Altered respiratory response to substance P and reduced NK1 receptor binding in the nucleus of the solitary tract of aged rats. Brain Res 1999;826: 139–142.

McDowell T, Gibbons R, Sutherland H, et al: Localization of a putative transcriptional regulator (ATRX) at pericentromeric heterochromatin and the short arms of acrocentric chromosomes. Proc Natl Acad Sci USA 1999;96:13983–13988.

McNulty JA, Tsai SY, Fox LM, et al: Neurotrophic effects of the pineal gland P. Role of non-neuronal cells in co-cultures of the pineal gland and superior cervical ganglia. J Pineal Res 1995;19:40–50.

Medina S, Del Rio M, Fernandez MD, et al: Changes with age in the modulation of natural killer activity of murine leukocytes by gastrin-releasing peptide, neuropeptide Y and sulfated cholecystokinin octapeptide. Neuropeptides 1998;32:549–555.

Melia KR, Ryabinin AE, Schroeder R, et al: Induction and habituation of immediate early gene expression in rat brain by acute and repeated restraint stress. J Neurosci 1994;14: 5929–5938.

Meßmer UK, Reed JC, Brüne B: Bcl-2 protects macrophages from nitric oxide-induced apoptosis. J Biol Chem 1996;47:20192–20197.

Meyerson M, Counter C, Eaton EN, et al: hEST2, the putative human telomerase catalytic subunit gene, is up-regulated in tumor cells and during immortalization. Cell 1997;90: 785–795.

Mhanna MJ, Dreshaj IA, Haxhui MA, Martin RJ: Mechanism for substance P-induced relaxation of pre-contracted airway smooth muscle during development. Am J Physiol 1999;276:L51–L56.

Middendorff R, Maronde E, Paust HJ, et al: Expression of C–type natriuretic peptide in the bovine pineal gland. J Neurochem 1996;67:517–524.

Miller MM, Bennett HJ, Billiar RB, et al: Estrogen, the ovary, and neurotransmitters: factors associated with aging. Exp Gerontol 1998;33:729–757.

Mita K, Tsuji H, Morimyo M, et al: The human gene encoding the largest subunit of RNA polymerase II. Gene 1995;159:285–286.

Mitchell PJ, Tijan R: Transcriptional regulation in mammalian cells by sequence-specific DNA-binding proteins. Science 1989;245:371–378.

Moller M, Ravault JP, Cozzi B: The chemical neuroanatomy of the mammalian pineal gland: neuropeptides. Neurochem Int 1996;28:23–33.

Morozov VG, Khavinson VKh: Cytomedins – a new class of the biological regulators of multicellular systems. Uspekhi Sovremennoi Biol 1983;96:339–352.

Morozov VG, Khavinson VKh: Peptide bioregulators (25 years of experimental and clinical study). St. Petersburg, Nauka, 1996.

Morozov VG, Khavinson VKh: Natural and synthetic thymic peptides as therapeutics for immune dysfunction. Int J Immunopharmacol 1997;19:501–505.

Morozov VG, Khavinson VKh, Malinin VV: Peptide Thymomimetics. St. Petersburg, Nauka, 2000.

Moujir F, Sanchez-Franco F, Santana C, et al: Immunoreactive levels of pineal arginine vasopressin change during the rat estrous cycle. J Pineal Res 1990;8:359–366.

Mouzaki A, Weil R, Muster L, Rungger D. Silencing and transactivation of the mouse interleikin-2 gene in *Xenopus* oocytes by proteins from resting and mitogen-induced primary T-lymphocytes. EMBO J 1991;10:1399–1406.

Mulchahey JJ, Kaskow JW, Plotsky PM, Hauger RL: Steroidal regulation of portal arginine-vasopressin levels in aged Fischer 344 rats. Brain Res 1999;822:243–245.

Mulligan T, Iranmanesh A, Kerzner R, et al: Two-week pulsatile gonadotropin releasing hormone infusion unmasks dual (hypothalamic and Leydig cell) defects in the healthy aging male gonadotropic axis. Eur J Endocrinol 1999;141:257–266.

Nandel FS, Malik N, Virdi M, Singh B: Designing of peptides with left handed helical structure by incorporating the unusual amino acids. Indian J Biochem Biophys 1999;36:195–203.

Odagiri Y, Uchida H, Hosokawa M, et al: Accelerated accumulation of somatic mutations in the senescence-accelerated mouse. Nat Genet 1998;19:116–117.

Ogryzko VV, Kotani T, Zhang X, et al: Histone-like TAFs within the PSAF histone acetilase complex. Cell 1998;94:35–44.

Oladehin A, Blatteis CM: Lipopolysaccharide-induced Fos expression in hypothalamic nuclei of neunatal rats. Neuroimmunorehabilitation 1995;2:282–289.

Olson M, Dundr M, Szebeni A: The nucleolus: An old factory with unexpected capabilities. Trends Cell Biol 2000;10:189–196.

Ouyang Q, Cicek G, Westendorp RG, et al: Reduced IFN-gamma production in elderly people following in vitro stimulation with influenza vaccine and endotoxin. Mech Age Dev 2000;121:131–137.

Pabo CO, Sauer RT: Transcription factors: Structural families and principles of DNA recognition. Ann Rev Biochem 1992;61:1053–1095.

Paltsev MA, Ivanov AA: Intercellular Interactions. Moscow, Meditsina, 1995.

Park MK, Kogo H, Kawashima S, Wakabayashi K: Characterization of gonadotropin-releasing hormone (GnRH)-immunoreactive protein in the rat pineal gland. J Neurosci Res 1995;41:386–393.

Parkinson G, Wilson C, Gunasekera A, et al: Structure of the CAP-DNA complex at 2.5 Å resolution: A complete picture of the protein-DNA interface. J Mol Biol 1996;60:395–408.

Pawelec G, Effros RB, Caruso C, et al: T cells and aging. Front Biosci 1999;4:D216–D269.

Pauling L: A molecular theory of general anesthesia. Science 1961;134:15–21.

Pekary AE, Hershman JM: Tumor necrosis factor, ceramide, transforming growth factor-beta1, and aging reduce Na^{+}/I^{-} symporter messenger ribonucleic acid levels in FRTL-5 cells. Endocrinology 1998;139:703–712.

Petrov RV, Mikhailova AA, Fomina LA, Stepanenko RN: Myelopeptides. Moscow, Nauka, 1997.
Petrov RV, Khaitov RM: Immune response and aging. Uspekhi Sovremennoi Biol 1975;79:111–127.
Pietrowsky R, Struben C, Molle M, Fehm HL: Brain potential changes after intranasal vs. intravenous administration of vasopressin: Evidence for a direct nose-brain effect in humans. Biol Psychiatry 1996a;39:332–340.
Pietrowsky R, Thiemann A, Kern W, et al: A nose-brain pathway for psychotropic peptides: Evidence from a brain evoked potential study with cholecystokinine. Psychoneuroendocrinology 1996b;21: 559–572.
Popov EM: Protein structural organization; in Protein problems, vol 3. Moscow, Nauka, 1997.
Raikhlin NT, Bukaeva IA, Smirnova EA, et al: Argyrophilic proteins from the nuclear organizer domain expression in thymocytes and epithelial cells of human thymus under conditions of combined cultivation under peptides vilon and epitalon action. Bull Exp Biol Med 2004;137:667–670.
Reiter RJ: The pineal gland and melatonin in relation to aging: A summary of the theories and of the data. Exp Gerontol 1995;30:199–212.
Rentzeperis D, Marky LA, Dwyer TJ, et al: Interaction of minor groove ligands to an AAATT/AATTT site correlation of thermodynamic characterization and solution structure. Biochemistry 1995;34: 2937–2945.
Rogers WO, Weaver CT, Kraus L, et al: Visualization of antigen-specific T cell activation and cytokine expression in vivo. J Immunol 1997;158:649–657.
Roudenok V, Kuhnel W, Rogov Y, Nerovnja A: Developmental changes in vasoactive intestinal polypeptide immunoreactivity in the human paravertebral ganglia. Anat Anz 1999;181:561–565.
Rosenfeld SV, Togo EF, Mikheev VS, et al: Effect of epitalon on the incidence of chromosome aberrations in senescence-accelerated mice. Bull Exp Biol Med 2002;133:274–276.
Rufer N, Dragowska W, Thornbury G, et al: Telomere length dynamics in human lymphocyte subpopulations measured by flow cytometry. Nat Biotechnol 1998;16:743–747.
Ryadnova IYu, Shataeva LK, Khavinson VKh: DNA-protein interaction studied in model systems. Polym Sci Ser [A] 2000;42:551–556.
Ryu DD, Nam DH: Recent progress in biomolecular engineering. Biotechnol Prog 2000;16:2–16.
Ryu S, Zhou S, Ladurner AG, Tijan R: The transcriptional cofactor complex CRSP is required for activity of the enhancer-binding protein Sp1. Nature 1999;397:446–450.
Sahu A, Kalra SP: Absence of increased neuropeptide Y neuronal activity before and during the luteinizing hormone (LH) surge may underlie the attenuated preovulatory LH surge in middle-aged rats. Endocrinology 1998;139:696–702.
Sazanov LA, Zaitsev SV: Effect of superlow doses (10^{-18}–10^{-14} *M*) of biologically active substances: General rules, features, and possible mechanisms. Biokhimiia 1992;57:1443–1460.
Schmallhausen II: Cybernetic Problems of Biology. Novosibirsk, Nauka SO, 1968.
Schondorf T, Kurbacher CM, Becker M, et al: Heterogeneity of proteinkinase C activity and PKC-zeta expression in clinical breast carcinomas. Clin Exp Med 2001;1:1–8.
Seeman NC, Rosenberg JM, Rich A: Sequence-specific recognition of double helical nucleic acids by proteins. Proc Natl Acad Sci USA 1976;73:804–808.
Sei J, Vitkovic L, Yokoyama MM: Cytokines in the central nervous system: Regulatory roles in neuronal function, cell death and repair. Neuroimmunomodulation 1995;2:121–133.
Senba E, Umemoto S, Kawai Y, Noguchi K: Differential expression of fos family and jun family mRNAs in the rat hypothalamo-pituitary-adrenal axis after immobilization stress. Brain Res Mol Brain Res 1994;24:283–294.
Shataeva LK, Khavinson VKh, Ryadnova IYu: Peptide Self-Regulation of Living Systems (Facts and Hypotheses). St. Petersburg, Nauka, 2003.
Shataeva LK, Ryadnova IYu, Khavinson VKh: Information value of olygopeptide blocks in regulatory peptides and proteins. Uspekhi Sovremennoi Biol 2002;122:282–289.
Sibarov DA, Kovalenko RI, Malinin VV, Khavinson VKh: Epitalon influences pineal secretion in stress-exposed rats in the daytime. Neuro Endocrinol Lett 2002a;23:452–454.
Sibarov DA, Kovalenko RI, Nozdrachev AD, et al: Effects of pineal gland peptides on the electric activity of pinealocytes in rats. Dokl Biol Sci 2002b;385:331–333.
Sirri V, Roussel P, Trere D, et al: Amount variability of total and individual Ag-NOR proteins in cells stimulated to proliferate. J Histochem Cytochem 1995;43:887–893.

Stepan H, Leitner E, Bader M, Walther T: Organ-specific mRNA distribution of C-type natriuretic peptide in neonatal and adult mice. Regul Pept 2000;95:81–85.
Sudakov KV: Individual Resistance to Emotional Stress. Moscow, 1998.
Suske G: The Sp-family of transcription factors. Gene 1999;238:291–300.
Swanson LW: Brain Maps: Computer Graphics Files. Amsterdam, Elsevier Science, 1992.
Takeda T: Senescence-accelerated mouse (SAM): A biogerontological resource in aging research. Neurobiol Aging 1999;20:105–110.
Tavernarakis N, Driscoll M: Caloric restriction and life span: A role for protein turnover? Mech Age Dev 2002;123:215–229.
Terwel D, Markerink M, Jolles J: Age–related changes in concentrations of vasopressin in the central nervous system and plasma of the male Wistar rat. Mech Age Dev 1992;65:127–136.
Tollet-Egnell P, Flores-Morales A, Stahlberg N, et al: Gene expression profile of the aging process in rat liver: normalizing effects of growth hormone replacement. Mol Endocrinol 2001;15: 308–318.
Tucker JD, Spruill MD, Ramsey MJ, et al: Frequency of spontaneous chromosome aberrations in mice: effects of age. Mutat Res 1999;425:135–141.
Unlap T, Jope PS: Dexamethazone attenuates kainate-induced AP-1 activation in rat brain. Molecular Brain Res 1994;24:275–282.
Van der Zee EA, Jansen K, Gerkema MP: Severe loss of vasopressin-immunoreactive cells in the suprachiasmatic nucleus of aging voles coincides with reduced circadian organization of running wheel activity. Brain Res 1999;816:572–579.
Van Holde KE: Chromatin. New York, Springer, 1988.
Verrecchio A, Germann MW, Schick BP, et al: Design of peptides with high affinities for heparin and endothelial cell proteoglycans. J Biol Chem 2000;275:7701–7707.
Vijg J: Somatic mutations and aging: A re-evaluation. Mutat Res 2000;447:117–135.
Wainer PK, Langridge R, Blaney JM, Kollman PA: Electrostatic potential molecular surfaces. Proc Nat Acad Sci USA 1982;79:3754–3758.
Wang X, Zhang R, Gong P, Gong S: A study on the specific binding of some drugs with DNA and DNA bases. Abstr 3rd Ann Life Spring Forum of Biosci, IT and Capitals, Dalian, China, 2004.
Weindruch R, Kayo T, Lee C-K, Prolla TA: Gene expression profiling of aging using DNA microarrays. Mech Age Dev 2002;123:177–193.
Weiss O, Jimenez-Montano MA, Herzel H: Information content of protein sequences. J Theoret Biol 2000;206:379–386.
Wenner M, Kawamura N, Ishikawa T: Reward linked to increased natural killer cell activity in rats. Neuroimmunomodulation 2000;7:1–5.
Wick V, Zubov D, Hagen G: Genomic organisation and promoter characterization of the gene encoding the human telomerase reverse transcriptase (hTERT). Gene 1999;232: 97–106.
Wilson HL, McFie PJ, Roesler WJ: Different transcription factor binding arrays modulate the cAMP responsivity of the phosphoenolpyruvate carboxykinase gene promoter. J Biol Chem 2002;277: 43895–43902.
Wolfe SA, Zhou P, Dotsch V, et al: Unusual Rel-like architecture in the DNA-binding domain of the transcription factor NFATc. Nature 1997;385:172–176.
Worth LL, Jia SF, An T, Kleinerman ES: ImmTher, a lipophilic disaccharide derivative of muramyl dipeptide, up-regulates specific monocyte cytokine genes and activates monocyte-mediated tumoricidal activity. Cancer Immunol Immunother 1999;48:312–320.
Wright WE, Shay JW: Historical claims and current interpretations of replicative aging. Nat Biotechnol 2002;20:682–688.
Yaseen NR, Maizel AL, Wang F, Sharma S: Comparative analysis of NFAT complex in human T and B lymphocytes. J Biol Chem 1993;268:14285–14293.
Yokoyama C, Sasaki K: Regional expression of c-fos-like immunoreactivity in rat cerebral cortex after stress: Restraint and intraperitoneal lipopolysaccharide. Brain Res 1999;816: 267–275.
Yu BP, Yang R: Critical evaluation of the free radical theory of aging: a proposal for the oxidative stress hypothesis. Ann NY Acad Sci 1996;786:1–11.
Zamyatnin AA: Dilatometry in Protein Solutions. Moscow, Nauka, 1973.
Zenger V: Principles of nucleic acids structural organization. Moscow, Mir, 1987.

Zhao H, Patra A, Yeh CC, et al: Effects of aging on growth factors gene and protein expression in the dorsal and ventral lobes of rat prostate. Biochem Biophys Res Commun 2002;292:482–491.
Zhou T, Chiang CM: The intronless and NFNF-less human TAF (II) 55 gene contains a functional initiator and downstream promoter element. J Biol Chem 2001;276: 25503–25511.

Subject Index

Amino acids
biophysical features in peptides 64–67
composition of regulatory peptides 71, 72, 74
structure-function relationships in regulatory peptides 67–71

Bioregulative therapy
effective concentrations of peptides 73
geroprotective peptides, *see specific peptides*
overview 9
prospects 88–92

Caloric restriction, effects on gene expression 6
Carnosine, structure 67–69
c-fos
hypothalamus expression in stressed rats
Vilon effects in emotional stress 30, 31
Epitalon and Vilon effects in handling stress 31–35
pineal gland expression, Epitalon effects in osmotic stress 35–37
Chromatin, peptide effects in blood lymphocytes 46–51
Chromosomal aberrations
Epitalon reduction studies
hypokinesia rat model 40, 41
senescence-accelerated mice 41–43
Vilon reduction in hypokinesia rat model 40, 41

Complementarity
DNA-peptide interactions
modeling 79–86
peculiarities 78, 79
intermolecular interactions and conformations of peptides 70
matrix complementarity principle 70
Cortagen
blood lymphocyte chromatin effects 47–51
sequence 9
Cytomedins
bioregulative therapy 9
effective concentrations of peptides 73
geroprotective peptides, *see specific peptides*
history of study 8
peculiarities of structure 74, 75

DNA microarray
caloric restriction effects on gene expression 6
overview of gene expression changes in aging 6, 11
peptide effects on gene expression in aging mice
Epitalon 11–20
Vilon 11–20
DNA-peptide interactions
amino acids, biophysical features in peptides 64–67
base specificity of binding 75

complementarity
modeling 79–86
peculiarities 78, 79
DNA sequence blocks 84, 85
informational interaction 86, 87
peptide blocks in transcription factors 75–77
DNA synthesis, Vilon effects
hepatocyte mitotic activity after partial hepatectomy in rats 62, 63
lymphocytes in aged rats exposed to radiation 60–62
thymus cells treated with nitrites 58–60

Epitalon
anticarcinogenic activity in Her2/neu transgenic mice 38, 39
blood lymphocyte chromatin effects 46–51
c-fos expression effects in stressed rats
hypothalamus in handling stress 31–35
pineal gland in osmotic stress 35–37
chromosomal aberration reduction
hypokinesia rat model 40, 41
senescence-accelerated mice 41–43
DNA interaction modeling 82–84, 86, 87
DNA microarray analysis of effects on gene expression in aging mice 11–20
interleukin-2 gene expression response
hypothalamus in rats 24–29
lymphocytes in mouse spleen 21–23
mitochondrial gene expression response 13
nucleolus organizer effects in human thymic cells 43–46
oncogene expression response 13, 19
sequence 9
telomerase effects
activity 53
cell growth and proliferation 54, 55, 57
cell treatment 52
expression 53
telomere length 53, 54
Fibroblasts
Epitalon effects on telomerase
activity 53
cell growth and proliferation 54, 55, 57
cell treatment 52
expression 53
telomere length 53, 54
peptide secretion in aging 5
Fos, *see c-fos*

Hepatocyte, Vilon effects on mitotic activity after partial hepatectomy in rats 62, 63
Her2/neu, Epitalon and Vilon anticarcinogenic activity in transgenic mice 38, 39
Hypothalamus
age-related changes 4
c-fos expression in stressed rats
Epitalon and Vilon effects in handling stress 31–35
Vilon effects in emotional stress 30, 31
interleukin-2 gene expression response to Epitalon and Vilon 24–29

Inflammation, pathology 89
Interleukin-2 (IL-2), Epitalon and Vilon effects on gene expression
hypothalamus in rats 24–29
lymphocytes in mouse spleen 21–23

Livagen
blood lymphocyte chromatin effects 47–51
DNA interaction modeling 82, 83
sequence 9
Lymphocytes
chromatin, peptide effects in blood cells 46–51
DNA synthesis, Vilon effects in aged rats exposed to radiation 60–62
interleukin-2 gene expression response to Epitalon and Vilon in mouse spleen 21–23

Matrix complementarity, principles 70

Nucleolus organizers
effects of Vilon and Epitalon in human thymic cells 43–46
lymphocyte chromatin effects of peptides 46–51

Peptide-DNA interactions, *see* DNA-peptide interactions
Pharmacogenomics, bioregulative therapy considerations 90, 91
Pineal gland
age-related changes 3
c-fos, Epitalon effects on expression in osmotic stress 35–37
peptide secretion 3, 4
Pituitary, peptide secretion in aging 6
Prostamax
blood lymphocyte chromatin effects 47–51
sequence 9
Prostate, aging 5

Retinalamin, DNA interactions 85

Sarcopenia, pathophysiology 7
Secondary structure, geometrical parameters in peptides 78, 79
Sister chromatid exchange, lymphocyte chromatin effects of peptides 47–51

Telomerase
aging theory 52
components 52
Epitalon effects
activity 53
cell growth and proliferation 54, 55, 57
cell treatment 52
expression 53
telomere length 53, 54
Thymogen
immune modulation 68, 69
structure 67–69
Thymus
age-related involution 2
DNA synthesis, Vilon effects in cells treated with nitrites 58–60
nucleolus organizer effects of Vilon and Epitalon in human cell culture 43–46
peptide secretion 2
Transcription factors
binding sequence specificity 75
peptide blocks 75–77, 85
Tryptophan, membrane interactions in peptides 69

Vilon
anticarcinogenic activity in Her2/neu transgenic mice 38, 39
blood lymphocyte chromatin effects 46–51
c-fos expression effects in stressed rat hypothalamus
emotional stress 30, 31
handling stress 31–35
chromosomal aberration reduction in hypokinesia model 40, 41
DNA interaction modeling 79–81
DNA microarray analysis of effects on gene expression in aging mice 11–20
DNA synthesis effects
hepatocyte mitotic activity after partial hepatectomy in rats 62, 63
lymphocytes in aged rats exposed to radiation 60–62
thymus cells treated with nitrites 58–60
interleukin-2 gene expression response
hypothalamus in rats 24–29
lymphocytes in mouse spleen 21–23
mitochondrial gene expression response 13
nucleolus organizer effects in human thymic cells 43–46
oncogene expression response 13, 19
sequence 9
structure 68, 69